SpringerBriefs in Mathematics

SpringerBriefs in Mathematics showcases expositions in all areas of mathematics and applied mathematics. Manuscripts presenting new results or a single new result in a classical field, new field, or an emerging topic, applications, or bridges between new results and already published works, are encouraged. The series is intended for mathematicians and applied mathematicians. All works are peer-reviewed to meet the highest standards of scientific literature.

Titles from this series are indexed by Web of Science, Mathematical Reviews, and zbMATH.

More information about this series at http://www.springer.com/series/10030

Kunihiko Kodaira

Theory of Algebraic Surfaces

Springer

Kunihiko Kodaira
University of Tokyo
Tokyo, Japan

Translated by
Kazuhiro Konno
Takatsuki, Japan

ISSN 2191-8198 ISSN 2191-8201 (electronic)
SpringerBriefs in Mathematics
ISBN 978-981-15-7379-8 ISBN 978-981-15-7380-4 (eBook)
https://doi.org/10.1007/978-981-15-7380-4

This Springer imprint is published by the registered company Springer Nature Singapore Pte Ltd.
The registered company address is: 152 Beach Road, #21-01/04 Gateway East, Singapore 189721, Singapore

Notes taken by Shigeho Yamashima

Kunihiko Kodaira at his home in Tokyo, 1990 © Springer Japan

Foreword

In the academic year 1967, Kunihiko Kodaira gave a course of lectures at the University of Tokyo on the theory of complex algebraic surfaces. The lecture notes were published in 1968 as Volume 20 in the series of "Seminary Notes" by the University of Tokyo. That was the copy of the handwritten manuscript in Japanese by Shigeho Yamashima, based on his beautiful notes reflecting faithfully the atmosphere of Kodaira's lectures. The present book is an English translation of that volume with slight modifications, correcting typos, etc. in the Japanese version. The readers are expected to have only the elementary prerequisites on complex manifolds as the background.

The book consists of two parts: Chaps. 1 and 2.

After stating the goal of the lecture in the Introduction, in Chap. 1, basic facts on algebraic surfaces are reviewed, touching upon divisors, linear systems, intersection theory, and the Riemann–Roch theorem. It provides an elegant introduction to the theory of algebraic surfaces covering some classical materials whose modern proofs first appeared in Kodaira's papers. Among others, one can find a concise analytic proof of Gorenstein's theorem for curves on a non-singular surface, which is a detailed explanation of the one given in Appendix I to "On Compact Complex Analytic Surfaces, I," Ann. of Math. 71 (1960). Another highlight is the elementary proof of Noether's formula for the arithmetic genus of an algebraic surface. Nowadays, the formula is known and treated as a special case of Hirzebruch's Riemann–Roch theorem. Kodaira's approach is based on the standard fact that, via generic projections, every algebraic surface can be obtained as the normalization of a surface with only ordinary singularities in the projective 3-space. However, unlike the other modern proofs, the argument does not rely on general facts, such as Porteus' formula, which require a separate treatment. It is self-contained and follows a classical line, using Lefschetz pencils, much more in the style of Noether's original proof.

The second part, Chap. 2, discusses the behaviors of the Pluri-canonical maps of algebraic surfaces of general type, as an application of the general theory provided in Chap. 1. It gives a detailed account of "Pluri-canonical Systems on Algebraic Surfaces of General Type," J. Math. Soc. Japan 20 (1968). The main tool is a vanishing criterion for the first cohomology groups, which is known as Mumford's vanishing theorem. It is shown by the analytic method in essentially the same way as in the proof of Kodaira's vanishing theorem. It gives us, for example, the pluri-genus formula for algebraic surfaces of general type. Then by means of various composition series for pluri-canonical divisors, one can discuss when the pluri-canonical system is free from base points and when the pluri-canonical map is birational onto its image. Such excellent results have been extended by Kodaira himself, Enrico Bombieri, and others. So, we can now answer affirmatively to a question stated in Remark 9 in the Introduction.

Takatsuki, Japan
April 2020

Kazuhiro Konno

Preface

This grew out of a course of lectures given by Professor Kunihiko Kodaira at the Department of Mathematics, Faculty of Science, the University of Tokyo, two hours per week from September 1967 to February 1968. The notes were taken and carefully arranged by Mr. Shigeho Yamashima. I would like to express a deep gratitude to him.

May 1968 Yukiyoshi Kawada

Introduction (Purposes and Known Results)

Throughout, we let S denote a non-singular projective algebraic surface defined over the field $\mathbb{C}$ of complex numbers, that is, it is a two-dimensional compact complex manifold embedded into a complex projective space $\mathbb{P}^N(\mathbb{C})$. In general, a two-dimensional complete algebraic variety without singular points is projective.

Notation 1

K: the canonical line bundle on S.
$mK = K \otimes \cdots \otimes K = K^{\otimes m}$: m-times tensor product of K.
$\mathscr{L}(mK)$: the linear space consisting of (global) holomorphic sections of mK.
$P_m = \dim_{\mathbb{C}} \mathscr{L}(mK)$: the m-genus of S.
$\{\varphi_0, \varphi_1, \ldots, \varphi_n\}$: a basis for $\mathscr{L}(mK)$ over $\mathbb{C}$.

$$\begin{array}{rccl} \Phi_{mK}\colon S & \longrightarrow & \mathbb{P}^n & \text{the rational map defined by } mK. \\ \Psi & & \Psi & \\ z & \rightsquigarrow & (\varphi_0(z), \varphi_1(z), \ldots, \varphi_n(z)) & \end{array}$$

c_1: the first Chern class of S.

c_1^2 can be considered as an integer via $H^4(S, \mathbb{Z}) \cong \mathbb{Z}$.

Assumption 2 *S contains no exceptional curves of the first kind. (An exceptional curve of the first kind is a non-singular rational curve with self-intersection number -1.)*

Then, S is said to be *relatively minimal.*

Assumption 3 *$P_2 > 0$ and $c_1^2 > 0$.*

Definition 4 When Assumptions 2 and 3 are satisfied, S is called a *minimal non-singular algebraic surface of general type.*

Remark 5

(i) If $P_m > 0$ for some m, then a relatively minimal model S is the minimal model.
(ii) In Šafarevič [8], the term *of fundamental type* is used for *of general type*.

We assume that S satisfies 2 and 3.

Problem 6 Study Φ_{mK}.

7 Known results

(a) D. Mumford [5]: There exists an integer $m_0(S)$ such that Φ_{mK} is a birational holomorphic map for any $m \geqq m_0(S)$.
(b) I.P. Šafarevič [8]: Φ_{9K} is a birational map (not referred to whether it is holomorphic or not).

The main purpose of the book is to prove the following:

Theorem 8 (Theorem 8.17) Φ_{mK} *is a birational holomorphic map when* $m \geqq 6$.

Remark 9 Can we relax the condition to $m \geqq 5$? There is an example for which Φ_{5K} is a birational holomorphic map, while Φ_{4K} is not birational.

In fact, (i) let $W \subset \mathbb{P}^3$ be the surface defined by $\sum_{i=0}^{3} z_i^5 = 0$. For $\varepsilon = \exp(\frac{2\pi \mathrm{i}}{5})$, put $g : (z_0, z_1, z_2.z_3) \to (z_0, \varepsilon z_1, \varepsilon^{-1} z_2, z_3)$ and $G = \{g^n \mid n = 0, 1, \dots, 4\}$. Then W/G has 5 singular points corresponding to $(1, 0, 0, \varepsilon^\nu)$ $(0 \leqq \nu \leqq 4)$. If S is its desingularization [1], then Φ_{4K} is not birational, but Φ_{5K} is a birational holomorphic map. (ii) When W is defined by $\sum_{i=0}^{3} z_i^6 = 0$, a similar construction gives us a surface S for which Φ_{4K} is birational.

Remark 10 If S satisfies Assumption 2 but not Assumption 3 (that is, either $P_2 = 0$ or $c_1^2 \leqq 0$), then S is one of the following five types of surfaces:

(a) the projective plane $\mathbb{P}^2$,
(b) a ruled surface,
(c) a K3 surface,
(d) an Abelian surface (two-dimensional complex torus),
(e) an elliptic surface.

Contents

Chapter 1
Fundamentals of Algebraic Surfaces

Abstract Fundamental properties of algebraic surfaces are collected and reviewed, such as curves on a surface, divisors and linear systems and the Riemann–Roch theorem. An elementary proof of Noether's formula is also given.

1.1 Exact Sequences

Notation 1.1

S:	a non-singular algebraic surface.
$C \subset S$:	an algebraic curve on S (assumed to be irreducible).
$\mu : \widetilde{C} \to C$:	the desingularization of C, that is, $\widetilde{C}$ is a non-singular model of C (unique up to isomorphisms) and μ is a birational holomorphic map.
$\mathcal{O} = \mathcal{O}_S$:	the sheaf of germs of holomorphic functions on S.
$\mathcal{O}(-C)$:	the sheaf of germs of holomorphic functions on S vanishing on C (this is a subsheaf of $\mathcal{O}$).

We put $\mathcal{O}_C = \mathcal{O}/\mathcal{O}(-C)$.

1.2 We study the structure of $\mathcal{O}_C$.

For $x \in S$, we denote by (w, z) a system of local coordinates of the center x on S. Then $\mathcal{O}_x = \mathcal{O}_{S,x} = \mathbb{C}\{w, z\}$, where $\mathbb{C}\{w, z\}$ denotes the ring of convergent power series in the indeterminates w, z with coefficients in $\mathbb{C}$.

In the following, we separately consider three cases (a), (b), and (c) divided according to the position of x for C:

(a) The case where $x \notin C$.
 Obviously, we have $(\mathcal{O}_C)_x = \{0\}$.

(b) The case where x is a simple point of C.
 As in the figure, we take a system of local coordinates (w, z) on S with the center x such that C is defined by $w = 0$ in a neighborhood of x. Then $\mathcal{O}(-C)_x = w\mathbb{C}\{w, z\}$ and, hence,

$$(\mathcal{O}_C)_x = \mathcal{O}_x/\mathcal{O}(-C)_x = \mathbb{C}\{w, z\}/w\mathbb{C}\{w, z\} \cong \mathbb{C}\{z\}.$$

K. Kodaira, *Theory of Algebraic Surfaces*, SpringerBriefs in Mathematics,
https://doi.org/10.1007/978-981-15-7380-4_1

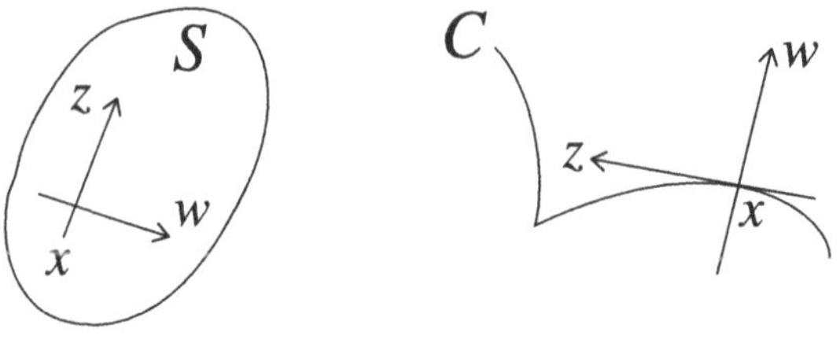

(c) The case where x is a singular point of C.

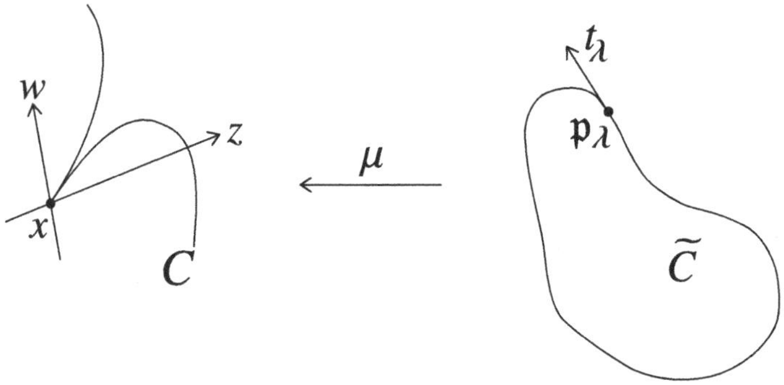

The inverse image $\mu^{-1}(x)$ consists of a finite number of points $\{\mathfrak{p}_1, \ldots, \mathfrak{p}_\lambda, \ldots, \mathfrak{p}_r\}$ on $\widetilde{C}$. We take a local parameter t_λ around $\mathfrak{p}_\lambda$ on $\widetilde{C}$ such that μ can be expressed as

$$\mu \colon t_\lambda \to (w, z) = (P_\lambda(t_\lambda), t_\lambda^{m_\lambda})$$

in a neighborhood of $\mathfrak{p}_\lambda$, where $P_\lambda(t_\lambda) \in t_\lambda^{m_\lambda}\mathbb{C}\{t_\lambda\}$. Let $C_\lambda = \{\mu(t_\lambda) \mid |t_\lambda| < \alpha\}$ be the irreducible analytic branch of C corresponding to $\mathfrak{p}_\lambda$. We write μ_λ for μ as above to specify the index (that is, μ_λ is the restriction of μ to a neighborhood of $\mathfrak{p}_\lambda$).

We define R_λ, R as follows:

$$R_\lambda(w, z) = \prod_{k=0}^{m_\lambda - 1} \left(w - P_\lambda(\varepsilon_\lambda^k z^{1/m_\lambda})\right) = w^{m_\lambda} + A_{\lambda,1}(z) w^{m_\lambda - 1} + \cdots + A_{\lambda, m_\lambda}(z),$$

$$R(w, z) = \prod_{\lambda=1}^{r} R_\lambda(w, z) = w^m + A_1(z) w^{m-1} + \cdots + A_m(z),$$

where $\varepsilon_\lambda = \exp(2\pi i/m_\lambda)$, $A_k(z) \in z^k\mathbb{C}\{z\}$ and $m = \sum_{\lambda=1}^r m_\lambda$. Then $C : R(w, z) = 0$ in a neighborhood of x. We have a homomorphism $\mu_\lambda^* : \mathscr{O} \to \mathbb{C}\{t_\lambda\} (=: \mathfrak{o}_\lambda)$ defined by

$$\mu_\lambda^* \colon \Phi(w, z) \mapsto (\mu_\lambda^* \Phi)(t_\lambda) = \Phi(P_\lambda(t_\lambda), t_\lambda^{m_\lambda}).$$

Put $\mathfrak{o} = \oplus_{\lambda=1}^{r} \mathfrak{o}_\lambda$. Then, we have the homomorphism $\mu^* : \mathscr{O}_x \to \mathfrak{o}$ defined by

$$\mu^* : \Phi \mapsto \mu_1^* \Phi + \cdots + \mu_\lambda^* \Phi + \cdots + \mu_r^* \Phi,$$

where $\Phi = \Phi(w, z) \in \mathscr{O}_x = \mathbb{C}\{w, z\}$. We obviously have $\mathscr{O}(-C)_x = \{\Phi \in \mathscr{O}_x \mid \mu^* \Phi = 0\}$ and, therefore,

$$(\mathscr{O}_C)_x = \mathscr{O}_x / \mathscr{O}(-C)_x = \mu^* \mathscr{O}_x \subset \mathfrak{o}.$$

So, it suffices to study $\mu^* \mathscr{O}_x$, which will be done later in 1.4.

Example 1.2.1 Here is an example in which C has a unique irreducible analytic branch at $x \in C$. Put $C = C_1$ in a neighborhood of x and $\mu = \mu_1 : t \mapsto (w, z) = (t^q, t^m)$, where $q > m > 0$, $(q, m) = 1$.

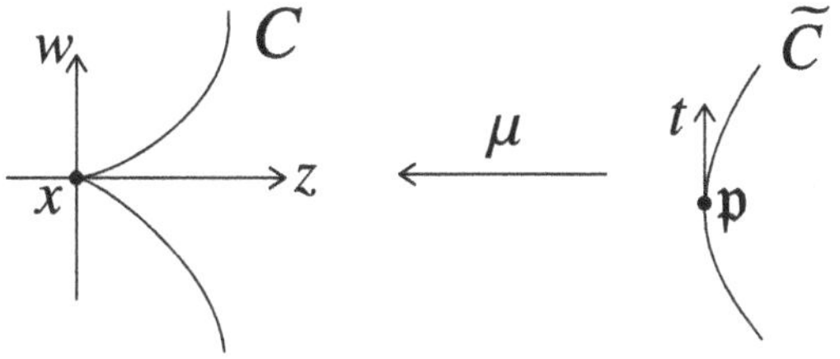

Since we have $(\mu^* \Phi)(t) = \sum_{h,k \geqq 0} a_{hk} t^{hq+mk}$ for $\Phi = \sum a_{hk} w^h z^k \in \mathscr{O}_x$, we obtain

$$\mu^* \mathscr{O}_x = \{\mu^* \Phi \mid \Phi \in \mathscr{O}_x\} = \{\varphi(t) \mid \varphi(t) = \sum_{\substack{n = hq + km \\ h, k \geqq 0}} b_n t^n\}.$$

If we put $m = 5$, $q = 7$ for example, then we have

$$\begin{aligned} \{hq + km\}_{h,k \geqq 0} &= \{7h + 5k\}_{h,k \geqq 0} \\ &= \{0, 5, 7, 10, 12, 14, 15, 17, 19, 20, 21, 22, 24, 25, 26, 27, \ldots\}. \end{aligned}$$

In this sequence, the number of elements from 0 to 22 is $12 = \frac{1}{2}(m-1)(q-1)$ and we have all integers greater than or equal to $24 = (m-1)(q-1)$. See Example 1.4.1.

Definition 1.3 In the situation of 1.2(c), we put

$$\sigma_\lambda \mathrm{d}t_\lambda = \frac{\mathrm{d}(t_\lambda^{m_\lambda})}{R_w(P_\lambda(t_\lambda), t_\lambda^{m_\lambda})} \quad \left(\text{where } R_w(w, z) = \frac{\partial R(w, z)}{\partial w}\right). \tag{1.3.1}$$

Then we can express σ_λ as

$$\sigma_\lambda = t_\lambda^{-c_\lambda}(a_{\lambda,0} + a_{\lambda,1}t_\lambda + \cdots) \qquad (a_{\lambda,0} \neq 0).$$

Since x is a singular point, c_λ is a positive integer. We put

$$\widetilde{\mathfrak{o}} = \bigoplus_{\lambda=1}^{r} t_\lambda^{c_\lambda}\mathfrak{o}_\lambda \quad \left(\subset \mathfrak{o} = \bigoplus_{\lambda=1}^{r} \mathfrak{o}_\lambda\right). \tag{1.3.2}$$

Remark 1.3.3 Even when we are in the situation of 1.2(b), we can consider 1.3.1. Since then $m_\lambda = 1$, $R(w, z) = w$, we get $\sigma_\lambda = 1$ and $c_\lambda = 0$. (In fact, $r = 1$.)

Theorem 1.4 (Gorenstein [3] and Rosenlicht [7])

(1) $\widetilde{\mathfrak{o}}$ *is a subring of* $\mu^*\mathcal{O}_X$*, that is,* $\widetilde{\mathfrak{o}} \subset \mu^*\mathcal{O}_X \subset \mathfrak{o}$.

(2) $\dim\,[\mathfrak{o}/\mu^*\mathcal{O}_X] = \dim\,[\mu^*\mathcal{O}_X/\tilde{\mathfrak{o}}] = \dfrac{1}{2}\sum_{\lambda=1}^{r} c_\lambda.$

Example 1.4.1 In the situation of Example 1.2.1, we have

$$R(w, z) = w^m - z^q, \quad P(t) = t^q$$

and, hence, $R_w = mw^{m-1}$ and $\sigma\,\mathrm{d}t = \dfrac{\mathrm{d}(t^m)}{mt^{q(m-1)}} = \dfrac{\mathrm{d}t}{t^{(m-1)(q-1)}}$. In particular, when $m = 5$ and $q = 7$,

$$\begin{cases} \mathfrak{o} \longleftrightarrow \mathbb{Z}^{\geqq 0} = \{0, 1, 2, 3, \ldots\} \\ \mu^*\mathcal{O}_X \longleftrightarrow \{7h + 5k\}_{h,k\geqq 0} \qquad (\text{see } 1.2.1) \\ \widetilde{\mathfrak{o}} \longleftrightarrow \{24, 25, \ldots\} \\ c = c_1 = (m-1)(q-1) = 4 \times 6 = 24. \end{cases}$$

In fact, we have $\mathfrak{o} = \mathbb{C}\{t\}$ and $\widetilde{\mathfrak{o}} = t^{24}\mathbb{C}\{t\}$. So, in this case Theorem 1.4 has been confirmed.

Problem 1.4.2 Generalize Theorem 1.4 to higher-dimensional cases.

1.5 From here to the end of 1.6, we prepare things for the proof of Theorem 1.4 which will be given in 1.7.

Let the situation be as in 1.2(c). We define $\mathfrak{f}_\lambda$, $\mathfrak{f}$, $\mathscr{R}$ as follows:

$$\mathfrak{f}_\lambda = \text{the quotient field of } \mathfrak{o}_\lambda = \{\varphi(t_\lambda) \mid \varphi(t_\lambda) = \sum_{n=m}^{+\infty} a_n t_\lambda^n\} = \mathbb{C}(\{t_\lambda\})$$

$$\text{(hereafter, } \mathbb{C}(\{z\}) = \text{the quotient field of } \mathbb{C}\{z\}),$$

$$\mathfrak{f} = \bigoplus_{\lambda=1}^{r} \mathfrak{f}_\lambda,$$

$$\mathscr{R} = \left\{ \frac{\Phi}{\Psi} \;\middle|\; \Phi, \Psi \in \mathscr{O}_x,\ \mu_\lambda^* \Psi \neq 0 \text{ for any } \lambda \right\}.$$

The homomorpism $\mu^* \colon \mathscr{O} \to \mathfrak{o}$ can be extended to the homomorphism $\tilde{\mu}^* \colon \mathscr{R} \to \mathfrak{f}$, $\tilde{\mu}^*|_{\mathscr{O}_x} = \mu^*$. In fact, it suffices to put

$$\tilde{\mu}^* \colon \frac{\Phi}{\Psi} \mapsto \sum_{\lambda=1}^{r} \frac{\mu_\lambda^* \Phi}{\mu_\lambda^* \Psi}.$$

Lemma 1.5.1 *$\tilde{\mu}^* \colon \mathscr{R} \to \mathfrak{f}$ is surjective. In fact, any $\xi \in \mathfrak{f}$ can be expressed in the form*

$$\xi = \tilde{\mu}^* F,$$
$$F = F(w, z) = F_0(z) w^{m-1} + F_1(z) w^{m-2} + \cdots + F_{m-1}(z),$$

where $F_k(z) \in \mathbb{C}(\{z\})$, $m = \sum_{\lambda=1}^{r} m_\lambda$. Moreover, such an F is uniquely determined.

Proof

(a) Surjectivity of $\tilde{\mu}^*$: We put $w_\lambda^* = P_\lambda(t_\lambda)$, $z_\lambda^* = t_\lambda^{m_\lambda}$ and $\mathfrak{f}_\lambda^* = \mathbb{C}(\{z_\lambda^*\})$ $(\subset \mathfrak{f}_\lambda)$. Then we obviously have $[\mathfrak{f}_\lambda : \mathfrak{f}_\lambda^*] = m_\lambda$. On the other hand, since $R_\lambda(w, z_\lambda^*) = w^{m_\lambda} + A_{\lambda 1}(z_\lambda^*) w^{m_\lambda - 1} + \cdots + A_{\lambda m_\lambda}(z_\lambda^*)$ $(A_{\lambda k}(z_\lambda^*) \in \mathfrak{f}_\lambda^*)$ is irreducible and $R_\lambda(w_\lambda^*, z_\lambda^*) = 0$, we have $[\mathfrak{f}_\lambda^*(w_\lambda^*) \colon \mathfrak{f}_\lambda^*] = m_\lambda$. Since we have $\mathfrak{f}_\lambda \subseteq \mathfrak{f}_\lambda^*(w_\lambda^*)$ and $[\mathfrak{f}_\lambda : \mathfrak{f}_\lambda^*] = m_\lambda$, we obtain

$$\mathfrak{f}_\lambda = \mathfrak{f}_\lambda^*(w_\lambda^*). \qquad (*)$$

Next, we put $S_\lambda = S_\lambda(w, z) = \prod_{\nu \neq \lambda} R_\nu(w, z)$ and $s_\lambda = \mu_\lambda^* S_\lambda$ $(\in \mathfrak{o}_\lambda)$. Express the given $\xi \in \mathfrak{f}$ as $\xi = \sum_{\lambda=1}^{r} \xi_\lambda$ $(\xi_\lambda \in \mathfrak{f}_\lambda)$. Noting that $s_\lambda \neq 0$, we put $\eta_\lambda = s_\lambda^{-1} \xi_\lambda$ $(\in \mathfrak{f}_\lambda)$. Then by $(*)$ we can write

$$\eta_\lambda = F_{\lambda 0}^* \cdot (w_\lambda^*)^{m_\lambda - 1} + \cdots + F_{\lambda m_\lambda - 1}^*,$$

where $F_{\lambda k}^* = F_{\lambda k}(z_\lambda^*)$ ($\in \mathfrak{f}_\lambda^*$) is a meromorphic function in z_λ^*. Noting that $F_{\lambda k}(z) \in \mathbb{C}(\{z\})$, if we put

$$F_\lambda(w, z) = F_{\lambda 0}(z)w^{m_\lambda - 1} + \cdots + F_{\lambda m_\lambda - 1}(z),$$

then $\eta_\lambda = \tilde{\mu}_\lambda^* F_\lambda$, where $\tilde{\mu}_\lambda^* : \mathscr{R} \to \mathfrak{f}_\lambda$ denotes the λ-th component of $\tilde{\mu}^*$. We put $F(w, z) = \sum_{\lambda=1}^{r} S_\lambda(w, z) F_\lambda(w, z)$. Then it is of the form

$$F(w, z) = F_0(z)w^{m-1} + F_1(z)w^{m-2} + \cdots + F_{m-1}(z)$$

as wished and we have $\tilde{\mu}^* F = \sum_{\lambda=1}^{r} s_\lambda \eta_\lambda = \sum_{\lambda=1}^{r} \xi_\lambda = \xi$.

(b) The unicity of F: We shall show that we have $F = 0$ if either $F = 0$ or F is a polynomial in w of degree $m - 1$ satisfying $\tilde{\mu}^* F = 0$ ($F \in \mathscr{R}$). Recall that

$$\begin{aligned} R_\lambda(w, z) &= \prod_{k=0}^{m_\lambda - 1} (w - P_\lambda(\varepsilon_\lambda^k z_\lambda^{1/m_\lambda})) \\ &= w^{m_\lambda} + A_{\lambda 1}(z)w^{m_\lambda - 1} + \cdots + A_{\lambda m_\lambda}(z) \qquad (\varepsilon_\lambda = \exp(\frac{2\pi \mathrm{i}}{m_\lambda})) \end{aligned}$$

is irreducible and $R_\lambda(P_\lambda(t_\lambda), t_\lambda^{m_\lambda}) = 0$. On the other hand, we have $F(P_\lambda(t_\lambda), t_\lambda^{m_\lambda}) = 0$. Therefore, $F(w, z) \equiv 0$ (R_λ) and, hence, $F(w, z) \equiv 0$ (R). However, since $R = \prod_{\lambda=1}^{r} R_\lambda = w^m + A_1(z)w^{m-1} + \cdots + A_m(z)$ ($A_k(z) \in z^k \mathbb{C}\{z\}$) is a polynomial of degree m in w, we get $F(w, z) = 0$.

□

Definition 1.5.2 For a given $\eta = \sum_{\lambda=1}^{r} \eta_\lambda \in \mathfrak{f}$, $\eta_\lambda = \eta_\lambda(t_\lambda) \in \mathfrak{f}_\lambda$, we put

$$\rho(\eta) = \frac{1}{2\pi \mathrm{i}} \sum_{\lambda=1}^{r} \oint \eta_\lambda(t_\lambda)\sigma_\lambda(t_\lambda)\, \mathrm{d}t_\lambda,$$

where $\oint$ means the integral along a sufficiently small circle "$|t_\lambda| =$ constant", and $\sigma_\lambda = \sigma_\lambda(t_\lambda)$ is as in 1.3.1.

Definition 1.5.3 We put $B_h(w, z) = w^h + A_1(z)w^{h-1} + \cdots + A_h(z)$ (recall: $R(w, z) = w^m + A_1(z)w^{m-1} + \cdots + A_m(z)$).

Remark 1.5.4 When $R(\zeta, z) = 0$,

$$\frac{R(w, z)}{w - \zeta} = w^{m-1} + B_1(\zeta, z)w^{m-2} + B_2(\zeta, z)w^{m-3} + \cdots + B_{m-1}(\zeta, z).$$

Proof We let $Q(w, z; \zeta)$ be the right hand side of the above equation. Then $Q = w^{m-1} + (\zeta + A_1)w^{m-2} + (\zeta^2 + A_1\zeta + A_2)w^{m-3} + \cdots + (\zeta^{m-1} + A_1\zeta^{m-2} + \cdots + A_{m-1})$. We have

$$\begin{aligned}(w - \zeta)Q &= w^m + (\zeta + A_1)w^{m-1} + (\zeta^2 + A_1\zeta + A_2)w^{m-2} + \cdots \\ &\quad + (\zeta^{m-1} + A_1\zeta^{m-2} + \cdots + A_{m-1})w \\ &\quad - \zeta w^{m-1} - (\zeta^2 + A_1\zeta)w^{m-2} - \cdots - (\zeta^m + A_1\zeta^{m-1} + \cdots + A_{m-1}\zeta) \\ &= w^m + A_1 w^{m-1} + \cdots + A_m(z) = R(w, z),\end{aligned}$$

because $-(\zeta^m + A_1\zeta^{m-1} + \cdots + A_{m-1}\zeta) = A_m$ by $R(\zeta, z) = 0$. □

Lemma 1.5.5 *F as in* 1.5.1 *is given by the following formula:*

$$\begin{cases} F(w, z) = F_0(z)w^{m-1} + F_1(z)w^{m-1} + \cdots + F_{m-1}(z) \\ F_h(z) = \sum_n F_{hn} z^n \\ F_{hn} = \rho(\xi \cdot \tilde{\mu}^*(z^{-n-1}B_h)) \quad (\textit{cf. } 1.5.2) \\ \quad = \dfrac{1}{2\pi \mathrm{i}} \sum_\lambda \oint \dfrac{\xi_\lambda(t_\lambda) B_h(P_\lambda(t_\lambda), t_\lambda^{m_\lambda}) m_\lambda}{\partial_w R(P_\lambda(t_\lambda), t_\lambda^{m_\lambda}) t_\lambda^{n m_\lambda + 1}} \, \mathrm{d}t_\lambda \end{cases}$$

Proof Because $R(w, z) = \prod_{\lambda=1}^{r} \prod_{k=0}^{m_\lambda - 1} (w - \zeta_{\lambda k})$, $\zeta_{\lambda k} = P_\lambda(\varepsilon_\lambda^k z_\lambda^{\frac{1}{m_\lambda}})$, we have

$$F(w, z) = \sum_\lambda \sum_k \frac{R(w, z)}{w - \zeta_{\lambda k}} \frac{F(\zeta_{\lambda k}, z)}{\partial_w R(\zeta_{\lambda k}, z)}$$

by Lagrange's interpolarion formula. By 1.5.4, however,

$$\frac{R(w, z)}{w - \zeta_{\lambda k}} = w^{m-1} + B_1(\zeta_{\lambda k}, z)w^{m-2} + \cdots + B_h(\zeta_{\lambda k}, z)w^{m-h-1} + \cdots.$$

Hence, $F_h(z) = \sum_\lambda \sum_k B_h(\zeta_{\lambda k}, z)\dfrac{F(\zeta_{\lambda k}, z)}{\partial_w R(\zeta_{\lambda k}, z)}$. On the other hand, since we have $\xi_\lambda(t_\lambda) = \mu_\lambda^* F = F(P_\lambda(t_\lambda), t_\lambda^{m_\lambda})$, we get $F(\zeta_{\lambda k}, z) = \xi_\lambda(\varepsilon_\lambda^k z^{\frac{1}{m_\lambda}})$. It follows that $F_h(z) = \sum_\lambda \sum_k B_h(\zeta_{\lambda k}, z)\dfrac{\xi_\lambda(\varepsilon_\lambda^k z^{\frac{1}{m_\lambda}})}{\partial_w R(\zeta_{\lambda k}, z)}$.

Now, if we put $\dfrac{B_h(P_\lambda(t_\lambda), t_\lambda^{m_\lambda})\xi_\lambda(t_\lambda)}{\partial_w R(P_\lambda(t_\lambda), t_\lambda^{m_\lambda})} = \sum_{\nu=\nu_0}^{+\infty} \gamma_\nu^{(\lambda)} t_\lambda^\nu$, then we have

$$\sum_k B_h(\zeta_{\lambda k}, z)\frac{\xi_\lambda(\varepsilon_\lambda^k z^{\frac{1}{m_\lambda}})}{\partial_w R(\zeta_{\lambda k}, z)} = \sum_\nu \gamma_\nu^{(\lambda)} \sum_{k=0}^{m_\lambda - 1} \varepsilon_\lambda^{k\nu} z^{\frac{\nu}{m_\lambda}} = \sum_n \gamma_{nm_\lambda}^{(\lambda)} m_\lambda z^n,$$

because

$$\sum_{k=0}^{m_\lambda-1} \varepsilon_\lambda^{k\nu} = \begin{cases} m_\lambda, & \nu \equiv 0\ (m_\lambda) \\ 0, & \nu \not\equiv 0\ (m_\lambda) \end{cases} \quad (\text{where } \varepsilon_\lambda = \exp(\tfrac{2\pi \mathrm{i}}{m_\lambda})).$$

Hence, $F_h(z) = \sum_n F_{hn} z^n = \sum_\lambda \sum_n \gamma_{nm_\lambda}^{(\lambda)} m_\lambda z^n$ and we have

$$F_{hn} = \sum_\lambda \gamma_{nm_\lambda}^{(\lambda)} m_\lambda = \sum_\lambda \frac{1}{2\pi \mathrm{i}} \oint \frac{B_h(P_\lambda(t_\lambda), t_\lambda^{m_\lambda})\xi_\lambda(t_\lambda) m_\lambda}{\partial_w R(P_\lambda(t_\lambda), t_\lambda^{m_\lambda}) t_\lambda^{nm_\lambda+1}}\, \mathrm{d}t_\lambda.$$

□

Theorem 1.6 *A necessary and sufficient condition for $\xi \in \mathfrak{f}$ to be in $\mu^*\mathcal{O}_x$ is that $\rho(\xi \cdot \varphi) = 0$ holds for all $\varphi \in \mu^*\mathcal{O}_x$.*

Proof

(i) Sufficiency: We take $\xi \in \mathfrak{f}$ satisfying $\rho(\xi \cdot \varphi) = 0$ for all $\varphi \in \mu^*\mathcal{O}_x$. Since, by 1.5.5, it is given by $\xi = \tilde{\mu}^* F$, $F_{hn} = \rho(\xi \cdot \tilde{\mu}^*(z^{-n-1} B_h))$ $(F \in \mathscr{R})$, we only have to show that $F \in \mathcal{O}_x = \mathbb{C}\{w, z\}$. That is, it suffices to show that $F_{hn} = 0$ for $n \leqq -1$. When $n \leqq -1$, however, we have $z^{-n-1} B_h(w, z) \in \mathcal{O}_x$ and it follows that $\tilde{\mu}^*(z^{-n-1} B_h) \in \tilde{\mu}^*\mathcal{O}_x \subset \mathfrak{o}$. Hence $F_{hn} = \rho(\xi \cdot \tilde{\mu}^*(z^{-n-1} B_h)) = 0$ by the hypothesis.

(ii) Necessity: ξ can be written as $\xi = \tilde{\mu}^* F$ with some $F \in \mathscr{R}$ as in 1.5.5. Now, we assume that $\xi \in \tilde{\mu}^*\mathcal{O}_x$. Hence we can express $\xi = \tilde{\mu}^*\Phi$ with some $\Phi \in \mathcal{O}_x$. Firstly, we shall show that $F \in \mathcal{O}_x$. Recall that $F = F_0(z) w^{m-1} + \cdots + F_{m-1}(z)$ $(F_h(z) \in \mathbb{C}(\{z\}))$. We assume that $F \notin \mathcal{O}_x$ and derive a contradiction.

Assume that $F_h(z) \notin \mathbb{C}\{z\}$ for some h. If we write $F_h(z) = F_{h,-a_h} z^{-a_h} + \cdots$ $(F_{h,-a_h} \neq 0)$ and put $\max\{a_0, a_1, \dots, a_h, \dots\} = b$, then

$$\begin{cases} z^b F(w, z) = G(w, z) \in \mathbb{C}\{w, z\}, \\ G(w, 0) \neq 0. \end{cases}$$

Since $\tilde{\mu}^*\Phi = \tilde{\mu}^*F$, we have $\tilde{\mu}^*(G - z^b\Phi) = \tilde{\mu}^*(z^bF - z^b\Phi) = 0$. It follows that $G - z^b\Phi \equiv 0\ (R)$, that is, we can write $G(w, z) - z^b\Phi(w, z) = Q(w, z)R(w, z)$ $(Q(w, z) \in \mathbb{C}\{w, z\})$. Noting that $R = w^m + \cdots + A_h(z)w^{m-h} + \cdots$ $(A_h(z) \in z^h\mathbb{C}\{z\})$, we get

$$G(w, 0) = Q(w, 0)w^m.$$

Since, however, $G(w, 0)$ is a polynomial in w of degree $\leqq m-1$ and $Q(w, 0) \neq 0$, this is impossible. Therefore, $F \in \mathscr{O}_x$. We infer from 1.5.5 that for $n \leqq -1$

$$F_{hn} = \rho(\xi \cdot z^{-n-1} B_h) = 0.$$

In particular, $F_{0,-1} = \rho(\xi) = 0$. We have shown that $\rho(\eta) = 0$ holds for any $\eta \in \mu^*\mathscr{O}_x$. We need to show that $\rho(\xi \cdot \varphi) = 0$ for any $\varphi \in \mu^*\mathscr{O}_x$. But, since we have assumed that $\xi \in \mu^*\mathscr{O}_x$, we have $\xi \cdot \varphi \in \mu^*\mathscr{O}_x$. Therefore, $\rho(\xi \cdot \varphi) = 0$. □

1.7 Here we show Theorem 1.4.

For any submodule $\mathfrak{m}$ of $\mathfrak{f}$, we define a submodule $\mathfrak{m}'$ of $\mathfrak{f}$ by

$$\mathfrak{m}' = \{\xi \in \mathfrak{f} \mid \rho(\xi \cdot \eta) = 0 \text{ for all } \eta \in \mathfrak{m}\}.$$

Recall that $\mathfrak{o} = \bigoplus_{\lambda=1}^{r} \mathfrak{o}_\lambda$ and $\sigma_\lambda(t_\lambda) = t_\lambda^{-c_\lambda}(a_{\lambda 0} + a_{\lambda 1}t_\lambda + \cdots)$ $(a_{\lambda 0} \neq 0)$.

(a) $\mathfrak{o}' = \bigoplus_{\lambda=1}^{r} t_\lambda^{c_\lambda}\mathfrak{o}_\lambda$. Hence $\mathfrak{o}' \subset \mathfrak{o}$.

This is because $\rho(\xi \cdot \eta) = \sum_\lambda \dfrac{1}{2\pi \mathrm{i}} \oint \xi_\lambda(t_\lambda)\eta_\lambda(t_\lambda)\sigma_\lambda(t_\lambda)\,\mathrm{d}t_\lambda$ and, therefore,

$$\begin{aligned} \rho(\xi \cdot \eta) = 0 \text{ for } \forall \eta \in \mathfrak{o} &\Leftrightarrow \oint \xi_\lambda(t_\lambda)\eta_\lambda(t_\lambda)\sigma_\lambda(t_\lambda)\,\mathrm{d}t_\lambda = 0 \text{ for } \forall \eta_\lambda \in \mathbb{C}\{t_\lambda\}, \forall\lambda \\ &\Leftrightarrow \xi_\lambda(t_\lambda)\eta_\lambda(t_\lambda)\sigma_\lambda(t_\lambda) \text{ is holomorphic for } \forall \eta_\lambda \in \mathbb{C}\{t_\lambda\}, \forall\lambda \\ &\Leftrightarrow \xi_\lambda(t_\lambda) \equiv 0\ (t_\lambda^{c_\lambda}), \forall\lambda. \end{aligned}$$

If $\xi, \eta \in \mathfrak{o}$, then $\rho(\xi \cdot \eta)$ is determined modulo $\mathfrak{o}'$ and, hence, we regard $\rho(\xi \cdot \eta)$ as a (non-degenerate) bilinear function over $\mathfrak{o}/\mathfrak{o}'$. Then, by duality, we have

(b) $\mathfrak{o}'' = \mathfrak{o}$.

In general, if $\mathfrak{o}' \subset \mathfrak{m} \subset \mathfrak{o}$, then we infer from (b) that

(c) $\mathfrak{o}' \subset \mathfrak{m}' \subset \mathfrak{o}$ and $\dim(\mathfrak{m}/\mathfrak{o}') = \dim(\mathfrak{o}/\mathfrak{m}')$.

Now, by 1.6, we have $(\mu^*\mathcal{O}_x)' = \mu^*\mathcal{O}_x$. Therefore, since $\mu^*\mathcal{O}_x \subset \mathfrak{o}$, we get $\mathfrak{o}' \subset (\mu^*\mathcal{O}_x)' = \mu^*\mathcal{O}_x$. This shows 1.4(1). Applying (c) to $\mathfrak{o}' \subset \mu^*\mathcal{O}_x \subset \mathfrak{o}$, we get

$$\dim\,[\mu^*\mathcal{O}_x/\mathfrak{o}'] = \dim\,[\mathfrak{o}/\mu^*\mathcal{O}_x] = \frac{1}{2}\dim\,[\mathfrak{o}/\mathfrak{o}'].$$

On the other hand, we infer from (a) that $\dim\,[\mathfrak{o}/\mathfrak{o}'] = \sum_{\lambda=1}^{r} c_\lambda$. We have shown 1.4(2).

1.8 For any singular point $x \in C$, we write

$$\mathfrak{o}' = \mathfrak{o}'_x, \quad \mathfrak{o} = \mathfrak{o}_x$$

for the notation in 1.7. Recall that $\mu^*\mathcal{O}_x \cong (\mathcal{O}_C)_x$ (cf. 1.2(c)). We denote this isomorphism by τ_x. If we apply the above notations to a simple point $x \in C$, we have $\mathfrak{o}'_x = \mu^*\mathcal{O}_x = \mathfrak{o}_x = \mathfrak{o}$ and, of course, $\tau_x : \mu^*\mathcal{O}_x \cong (\mathcal{O}_C)_x$.

Definition 1.8.1 We denote by $\mathcal{O}'$ the subsheaf of $\mathcal{O}_C$ whose stalk over $x \in C$ is given by

$$\mathcal{O}'_x = \tau_x(\mathfrak{o}'_x)\ (\subseteq (\mathcal{O}_C)_x)\ (\text{recall } \mathfrak{o}'_x \subset \mu^*\mathcal{O}_x).$$

We define $\mathcal{M} = \mathcal{O}/\mathcal{O}'$. Then

$$0 \to \mathcal{O}' \to \mathcal{O}_C \to \mathcal{M} \to 0 \qquad \text{(exact)}.$$

1.8.2 For $x \in C$, $\mathcal{M}_x \cong \mu^*\mathcal{O}_x/\mathfrak{o}'_x$ is a finite-dimensional vector space. In fact, we have

$$\dim\,\mathcal{M}_x = \begin{cases} 0 & \cdots\ x : \text{simple point}, \\ \dfrac{1}{2}c_x & \cdots\ x : \text{singular point}, \end{cases}$$

where we put $c_x = \sum_{\mathfrak{p}_\lambda \in \mu^{-1}(x)} c_\lambda$.

1.9 Let F be a complex line bundle over S.

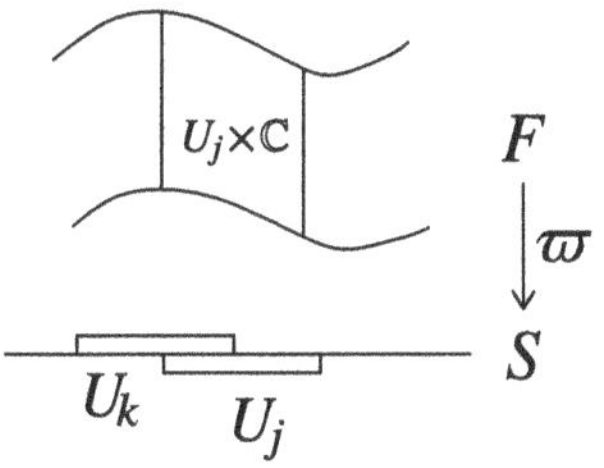

Then F can be described as follows:

(a) A surjective holomorphic map $\varpi : F \to S$ (called the canonical projection).
(b) A sufficiently fine open covering $S = \bigcup_j U_j$ such that $\varpi^{-1}(U_j) = U_j \times \mathbb{C}$.
(c) $U_j \times \mathbb{C}$ and $U_k \times \mathbb{C}$ are glued by the rule: for $z \in U_j \cap U_k$, $(z, \zeta_j) \in U_j \times \mathbb{C}$ and $(z, \zeta_k) \in U_k \times \mathbb{C}$,

$$(z, \zeta_j) = (z, \zeta_k) \Longleftrightarrow \zeta_j = f_{jk}(z)\zeta_k,$$

where f_{jk} is a (non-vanishing) holomorphic function defined on $U_j \cap U_k \neq \emptyset$ and we have $f_{ik} = f_{ij} f_{jk}$ on $U_i \cap U_j \cap U_k$. We call ζ_j the fibre coordinate of (z, ζ_j) over U_j.

1.9.1 By $\mathscr{O}^*$, we denote the multiplicative subsheaf of $\mathscr{O}$ such that $\Gamma(U, \mathscr{O}^*)$ consists of all invertible elements of $\Gamma(U, \mathscr{O})$. The 1-cocycle $\{f_{ij}\}$ as above determines an element of $H^1(S, \mathscr{O}^*)$, the isomorphism class of F. It is obvious that $H^1(S, \mathscr{O}^*)$ becomes an abelian group with product $F_1 \otimes F_2$. We regard $H^1(S, \mathscr{O}^*)$ as an additive group and write $F_1 + F_2$ instead of $F_1 \otimes F_2$.

1.9.2 A holomorphic section φ of F over S is a holomorphic map from S to F such that the composite $\varpi \circ \varphi$ is the identity map. So we can express it as:

$$\begin{cases} \varphi \colon z \to \varphi(z) = (z, \varphi_j(z)) \text{ (over } U_j), \\ \qquad \varphi_j(z) = f_{jk}(z)\varphi_k(z) \text{ on } U_j \cap U_k. \end{cases}$$

That is, φ can be identified with a collection $\{\varphi_j(z)\}$ of holomorphic functions $\varphi_j(z)$ on U_j satisfying $\varphi_j(z) = f_{jk}(z)\varphi_k(z)$.

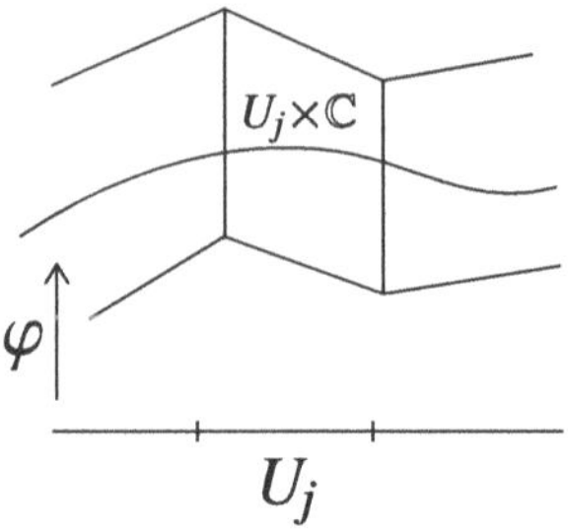

Definition 1.9.3 A collection $\varphi = \{\varphi_j(z)\}$ of meromorphic functions $\varphi_j(z)$ on U_j is called a *meromorphic section* of F if $\varphi_j(z) = f_{jk}(z)\varphi_k(z)$ holds on $U_j \cap U_k$.

1.10 Let F and C be a complex line bundle and a curve on S, respectively. We define sheaves as in the following (a), (b), and (c):

(a) $\mathcal{O}(F)$ = the sheaf of germs of holomorphic sections of F (a sheaf over S).
(b) $\mathcal{O}(F - C) = \{\varphi \in \mathcal{O}(F) \mid \varphi(z) = (z, 0) \text{ for } z \in C\}$.
(c) $\mathcal{O}(F)_C = \mathcal{O}(F)/\mathcal{O}(F - C)$: the restriction to C.

Therefore, $\mathcal{O}(F)_C = \mathcal{O}_C \otimes_{\mathcal{O}} \mathcal{O}(F) = \mathcal{O}_C \otimes F$. In fact:
$(\mathcal{O}_C|U_j = \tau^{-1}(U_j))$
$\mathcal{O}_C \otimes F = \bigcup_j \mathcal{O}_C|U_j$ is patched up by the following rule:

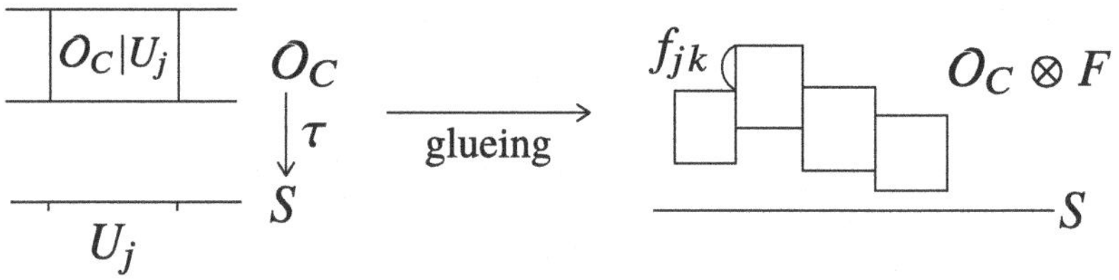

$$\begin{cases} \text{For } z \in U_j \cap U_k, \varphi_j \in (\mathcal{O}_C|U_j)_z \text{ and } \varphi_k \in (\mathcal{O}_C|U_k)_z, \\ \varphi_j = \varphi_k \iff \varphi_j = f_{jk}\varphi_k. \end{cases}$$

1.11 $\mathcal{O}'(F) = \mathcal{O}' \otimes F$ (cf. 1.8.1).

Then we have the following exact sequences (a), (b):

(a) $0 \to \mathcal{O}'(F) \to \mathcal{O}(F)_C \to \mathcal{M} \to 0$,
(b) $0 \to \mathcal{O}(F - C) \to \mathcal{O}(F) \to \mathcal{O}(F)_C \to 0$.

As to (a), we infer from 1.8.1 that the sequence $0 \to \mathcal{O}' \otimes F \to \mathcal{O}_C \otimes F \to \mathcal{M} \otimes F \to 0$ is exact and, since the sheaf $\mathcal{M}$ over C has non-zero stalks only over the singular points of C, we have $\mathcal{M} \otimes F \cong \mathcal{M}$; (b) is clear.

Definition 1.12 Let $\mu\colon \widetilde{C} \to C \subset S$ denote the desingularization of a curve C and let F be a complex line bundle over S. We denote by $\mu^* F$ the line bundle over $\widetilde{C}$ induced from F by μ, and by $\mathscr{O}(\mu^* F)$ the sheaf over $\widetilde{C}$ of germs of holomorphic sections of $\mu^* F$.

Definition 1.13 The following divisor $\mathfrak{c}$ on $\widetilde{C}$ is called the *conductor* on C:

$$\mathfrak{c} = \sum_{x \in \mathrm{Sing}(C)} \sum_{\mathfrak{p}_\lambda \in \mu^{-1}(x)} c_\lambda \mathfrak{p}_\lambda \qquad \text{(see 1.3 for the symbols).}$$

Then we put

$$\mathscr{O}(\mu^* F - \mathfrak{c}) = \{\eta \in \mathscr{O}(\mu^* F) \mid \eta \equiv 0 \ (\mathfrak{c})\}.$$

Lemma 1.13.1 $\mu_* \mathscr{O}(\mu^* F - \mathfrak{c}) \overset{\cong}{\longleftarrow} \mathscr{O}'(F)$.

Proof The natural homomorphism $\tau\colon \mathscr{O}'(F) \to \mathscr{O}(\mu^* F - \mathfrak{c})$ is given by $\tau\colon \varphi(w, z) \to (\mu^* \varphi)(t)$. It suffices to show that it induces the isomorphism on each stalk over any point on C. It is clear at simple points on C. So, let us consider a singular point $x \in C$. We employ the notation in 1.2. If we put $\mu^{-1}(x) = \{\mathfrak{p}_1, \ldots, \mathfrak{p}_\lambda, \ldots, \mathfrak{p}_r\}$, then $\mu_* \mathscr{O}(\mu^* F - \mathfrak{c})_x \cong \bigoplus_{\lambda=1}^{r} \mathscr{O}(\mu^* F - \mathfrak{c})_{\mathfrak{p}_\lambda}$. On the other hand, we infer from 1.13, 1.7, and 1.8 that there are canonical isomorphisms $\mathscr{O}(\mu^* F - \mathfrak{c})_{\mathfrak{p}_\lambda} \cong t_\lambda^{c_\lambda} \mathfrak{o}_\lambda$ and $\mathscr{O}'(F)_x \cong \mathscr{O}'_x \cong \bigoplus t_\lambda^{c_\lambda} \mathfrak{o}_\lambda$. Then, by the nature of the map μ, we get $\tau_x : \mathscr{O}'(F)_x \overset{\cong}{\to} \mu_*(\mathscr{O}(\mu^* F - \mathfrak{c}))_x$. □

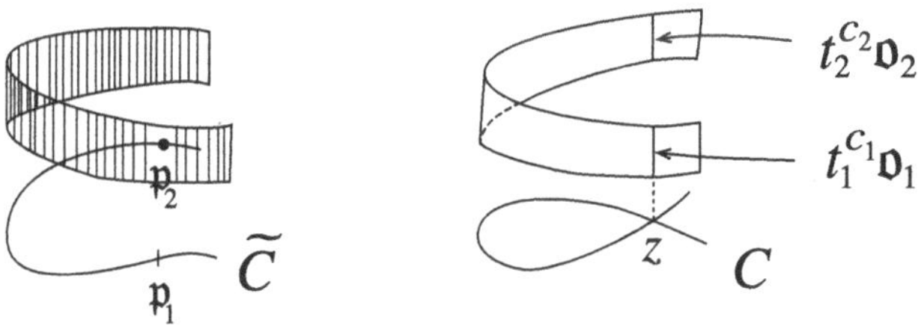

Theorem 1.14 *The following sequences are exact:*

$0 \to \mu_* \mathscr{O}(\mu^* F - \mathfrak{c}) \to \mathscr{O}(F)_C \to \mathscr{M} \to 0.$
$0 \to \mathscr{O}(F - C) \to \mathscr{O}(F) \to \mathscr{O}(F)_C \to 0.$

Proof The former follows from 1.11 and 1.13. The latter is obvious. □

We remark that $\mathscr{O}(F_C) = \mathscr{O}(F)_C$.

Theorem 1.15 *The following sequences are exact:*

$$\begin{aligned} 0 &\to H^0(S, \mathcal{O}(F - C)) \to H^0(S, \mathcal{O}(F)) \to H^0(C, \mathcal{O}(F)_C) \\ &\to H^1(S, \mathcal{O}(F - C)) \to H^1(S, \mathcal{O}(F)) \to H^1(C, \mathcal{O}(F)_C) \\ &\to H^2(S, \mathcal{O}(F - C)) \to H^2(S, \mathcal{O}(F)) \to 0. \end{aligned}$$

$$\begin{aligned} 0 &\to H^0(\widetilde{C}, \mathcal{O}(\mu^* F - \mathfrak{c})) \to H^0(C, \mathcal{O}(F_C)) \to H^0(C, \mathcal{M}) \\ &\to H^1(\widetilde{C}, \mathcal{O}(\mu^* F - \mathfrak{c})) \to H^1(C, \mathcal{O}(F_C)) \to 0. \end{aligned}$$

Moreover,

$$\dim H^0(C, \mathcal{M}) = \frac{1}{2} \sum_{x \in \mathrm{Sing}(C)} c_x \qquad (c_x = \sum_{\mathfrak{p}_\lambda \in \mu^{-1}(x)} c_\lambda).$$

Proof This follows from 1.14. We also remark that $H^\nu(\widetilde{C}, \mathcal{O}(\mu^* F - \mathfrak{c})) \cong H^\nu(C, \mathcal{O}'(F))$ holds by 1.13.1. The last equality follows from 1.8.2. □

1.2 Divisors and Linear Systems

2.1 Let C be an irreducible curve on S. We shall define the line bundle $[C]$ over S. For any point $x \in C$, if we take a sufficiently small neighborhood U of $x \in S$, then we can write $C \cap U = \{R(w, z) = 0\}$. Here, (w, z) is a system of local coordinates of S with the center x, $R(z) = R(z_1, z_2) = z_2^m + A_1(z_1)z_2^{m-1} + \cdots + A_m(z_1)$ is a holomorphic function such that $R(z) = 0$ is the minimal local equation (1.2), that is, for $y \in C \cap U$ and $\Phi \in \mathcal{O}_y$, if $\Phi(z) = 0$ holds for any $z \in C$ in a neighborhood of y, then $\Phi \equiv 0$ (R) in $\mathcal{O}_y$.

Let $S = \bigcup_j U_j$ be an open covering such that the minimal local equation of C on U_j is $R_j(z) = 0$. Then $f_{jk} = \frac{R_j(z)}{R_k(z)}$ is a non-vanishing holomorphic function on $U_j \cap U_k$. We of course have $f_{ik} = f_{ij} f_{jk}$ on $U_i \cap U_j \cap U_k$ and $\{f_{jk}\}$ defines a complex line bundle F.

Definition 2.1.1 We write $F = [C]$.

Since $R_j = f_{jk} R_k$ on $U_j \cap U_k$, $\varphi\colon z \to \varphi(z) = (z, R_j(z))$ is a holomorphic section of F, and $C = \{z \in S \mid \varphi(z) = (z, 0)\}$.

Definition 2.1.2 We write $C = (\varphi)$ and call it the divisor of φ.

2.2 We extend the above discussion to general divisors $D = \sum_{\nu=1}^r m_\nu C_\nu$. ($m_\nu \in \mathbb{Z}$, C_ν is an irreducible curve on S.)

Definition 2.2.1 $[D] = \sum_{\nu=1}^{r} m_\nu [C_\nu]$.

In fact, let $S = \bigcup_j U_j$ be an open covering such that the minimal local equation of C_ν on U_j is $R_{\nu j}(z) = 0$ and put $f_{jk} = \prod_{\nu=1}^{r} \left(\frac{R_{\nu j}}{R_{\nu k}}\right)^{m_\nu}$. Then $[D]$ is determined by $\{f_{jk}\}$ and $\varphi\colon z \to \varphi(z) = \left(z, \prod_{\nu=1}^{r} R_{\nu j}(z)^{m_\nu}\right)$ is a meromorphic section of $[D]$; we write $D = (\varphi)$ and call it the divisor of φ. Conversely, given a (not identically zero) meromorphic section φ of a line bundle F, it is immediate that we have a unique divisor D satisfying $F = [D]$ and $D = (\varphi)$.

Definition 2.2.2 When $[D] = 0$, where 0 denotes the trivial line bundle, that is, one defined by $f_{jk}(z) = 1$), we say that D is *linearly equivalent* to 0 and write $D \approx 0$. (Two divisors D_1 and D_2 are linearly equivalent, $D_1 \approx D_2$, if $D_1 - D_2 \approx 0$.)

2.2.3 The necessary and sufficient condition that $D \approx 0$ is that there exists a meromorphic function f on S satisfying $D = (f)$.

Proof We can write $D = (\varphi)$ with a meromorphic section $\varphi(z) = (z, \varphi_j)$ of $[D]$. If $D \approx 0$, then $[D] = 0$. Hence we have $\varphi_j(z) = \varphi_k(z)$ on $U_j \cap U_k$, that is, there exists a meromorphic function f on S such that $f(z) = \varphi_j(z)$ on U_j. Then $D = (f)$. The converse is clear. □

2.3 Let F be a line bundle over S. Then $H^0(S, \mathcal{O}(F))$ is the space of holomorphic sections of F.

Definition 2.3.1 We call $|F| = \{(\varphi) \mid \varphi \in H^0(S, \mathcal{O}(F)),\ \varphi \neq 0\}$ the *complete linear system* determined by F. (Here, $\varphi \neq 0$ means that φ is not identically zero.) For any $D \in |F|$, we have $D \geqq 0$ (see Remark below for the meaning of $\geqq$).

Remark 2.3.2 For $D = \sum_{\nu=1}^{r} m_\nu C_\nu$,

$$\begin{cases} \text{when } m_\nu > 0,\ D \text{ is said to be } \textit{positive and we write } D > 0, \\ \text{when } m_\nu \geqq 0,\ D \text{ is said to be } \textit{non-negative} \text{ (or } \textit{effective) and we write } D \geqq 0. \end{cases}$$

Definition 2.3.3 We put $|X| = |[X]|$ for a divisor X.

Proposition 2.3.4 $|X| = \{D \mid D \approx X,\ D \geqq 0\}$.

Proof Clear. □

Definition 2.4 Let D be a divisor and F a line bundle on S. We define the sheaf $\mathcal{O}(F - D)$ over S as follows:

(a) The case $D > 0$.

We take a holomorphic section $\varphi(z) = (z, \varphi_j(z))$ of F satisfying $D = (\varphi)$ and let $\mathcal{O}(F - D)$ be the subsheaf of $\mathcal{O}(F)$ determined by the following condition (obviously, it does not depend on the choice of φ):

The stalk over $x \in S$ is

$$\begin{aligned}\mathcal{O}(F-D)_x &= \{\psi \in \mathcal{O}(F)_x \mid \psi/\varphi_j \in \mathcal{O}_x\} \\ &= \{\psi \in \mathcal{O}(F)_x \mid \psi \equiv 0\ (D)\}.\end{aligned}$$

(b) General case.

We take a meromorphic section $\varphi(z) = (z, \varphi_j(z))$ of F satisfying $D = (\varphi)$. $\mathcal{O}(F-D)$ consists of meromorphic sections $\psi(z) = (z, \psi_j(z))$ of F satisfying $\psi_j/\varphi_j \in \mathcal{O}_x$ for $x \in S$.

Remark In the above argument, the expressions e.g., a section $\varphi(z) = (z, \varphi_j(z))$, are considered after taking a suitable open covering $S = \bigcup_j U_j$ as in 2.1. We apply the same rule in what follows.

Proposition 2.4.1 *There is a natural isomorphism $\mathcal{O}(F-D) \cong \mathcal{O}(F-[D])$.*

Proof We employ the notation in 2.4(b). We shall see that there is a natural isomorphism between sections over any open subset of S. In the following, we consider over an arbitrary fixed open subset of S. For any section $\psi\colon z \to \psi(z, \psi_j(z))$ of $\mathcal{O}(F-D)$, we put $\eta_j = \psi_j/\varphi_j$. Then $\eta\colon z \to (z, \eta_j(z))$ is a section of $\mathcal{O}(F-[D]) = \mathcal{O}(F \otimes [D]^{-1})$, because $\psi_j = f_{jk}\psi_k$, ψ_j/φ_j is holomorphic on U_j and $\eta_j/\eta_k = (\psi_j/\psi_k)(\varphi_k/\varphi_j) = f_{jk}\cdot g_{jk}^{-1}$ $(=: h_{jk})$ (where we put $g_{jk} = \varphi_j/\varphi_k$; $[D]$ is defined by $\{g_{jk}\}$). Therefore, $\{h_{jk}\}$ defines $F \otimes [D]^{-1}$, since $\eta_j = h_{jk}\eta_k$, we see that η is a holomorphic section of $F \otimes [D]^{-1}$. It is obvious that the map $\psi \to \eta$ gives us an isomorphism. □

1.3 Intersection Multiplicities and the Adjunction Formula

3.1 Let $(S, \mathcal{O})$ be a non-singular algebraic surface and $\mathcal{O}^*$ the multiplicative sheaf of germs of non-vanishing holomorphic functions (1.9.1). Then we have the following exact sequence:

$$0 \to \mathbb{Z} \to \mathcal{O} \xrightarrow{\tau} \mathcal{O}^* \to 0 \quad (\text{where } \tau\colon f \to e^{2\pi i f}),$$

and hence

$$\begin{array}{ccccc} \cdots \to & H^1(S, \mathcal{O}^*) & \xrightarrow{c} & H^2(S, \mathbb{Z}) & \to \cdots \\ & \Psi & & \Psi & \\ & F & \rightsquigarrow & c(F) & \end{array}$$

In fact, for a line bundle F, $c(F)$ denotes the Chern class of F. We look at this correspondence concretely:

If F is given by the 1-cocycle $\{f_{jk}\}$ for a covering $S = \bigcup_j U_j$ (where f_{jk} is a non-vanishing holomorphic function on $U_j \cap U_k$) and $H^2(S, \mathbb{Z}) \ni c(F) = \{c_{ijk}\}$ ($c_{ijk} \in \mathbb{Z}$), then it is easy to see that

$$c_{ijk} = \frac{1}{2\pi \mathrm{i}}\{\log f_{ij}(z) + \log f_{jk}(z) + \log f_{ki}(z)\}.$$

3.2 Let $F, G (\in H^1(S, \mathscr{O}^*))$ be line bundles and C, D divisors on S. We denote the value $(c(F)c(G))[S] \in \mathbb{Z}$ of $c(F) \cdot c(G) \in H^4(S, \mathbb{Z})$ on the fundamental 4-cycle S also by the same symbol $c(F) \cdot c(G)$.

Definition 3.2.1 $F \cdot G = c(F) \cdot c(G)$,
$C \cdot D = c([C]) \cdot c([D])$,
$F \cdot D = c(F) \cdot c([D])$.

Proposition 3.3 *Let $(R, \mathscr{O}_R)$ be a compact Riemann surface. Then the sheaf $\mathscr{O}_R^*$ can be defined as before and there is an exact sequence*

$$\begin{array}{ccccc} \cdots \to H^1(R, \mathscr{O}_R^*) & \xrightarrow{c} & H^2(R, \mathbb{Z}) \to, \\ \Psi & & \Psi \\ \mathfrak{f} & \rightsquigarrow & c(\mathfrak{f}) \end{array}$$

where $\mathfrak{f}$ denotes a line bundle over R. If $\mathfrak{d} = \sum_\nu m_\nu \mathfrak{p}_\nu$ ($m_\nu \in \mathbb{Z}, \mathfrak{p}_\nu \in R$) is a divisor on R, then

$$c([\mathfrak{d}]) = \sum_\nu m_\nu \ (= \deg \mathfrak{d}).$$

Proof We take a suitable open covering $R = \bigcup_j U_j$ and put $c([\mathfrak{d}]) = \{c_{ijk}\} \in H^2(R, \mathbb{Z})$ (an abbreviation of the representative $\{c_{ijk}\} \in Z^2(\mathscr{U}, \mathbb{Z})$, $\mathscr{U} = (U_j)$). Of course, we may think $\{c_{ijk}\} \in H^2(R, \mathbb{R})$. By de Rham's theorem, there is a 2-form ξ on R corresponding to $\{c_{ijk}\}$. In the following, we shall write ξ explicitly. Firstly, there exists a 1-cochain $\{\mu_{jk}\}$ (where μ_{ik} is a C^∞ function on $U_i \cap U_k$) such that $c_{ijk} = \mu_{ij} + \mu_{jk} + \mu_{ki}$ and $\mathrm{d}\mu_{ij} + \mathrm{d}\mu_{jk} + \mathrm{d}\mu_{ki} = 0$. $\{\mathrm{d}\mu_{jk}\}$ is the coboundary of a 0-cochain $\{\sigma_j\}$ (where σ_i is a C^∞ 1-form on U_i). That is, we have $\mathrm{d}\mu_{ij} = \sigma_j - \sigma_i$. Therefore, $\mathrm{d}\sigma_i = \mathrm{d}\sigma_j$ on $U_i \cap U_j$. This implies that $\xi = \mathrm{d}\sigma_i = \mathrm{d}\sigma_j = \cdots$ is a 2-form on R. It is clear that $c([\mathfrak{d}]) = \{c_{ijk}\} \leftarrow \xi$ (not necessarily unique, but the meaning may be clear). We also have the correspondence $\{c_{ijk}\} \leftrightarrow c \in \mathbb{Z}$, because $H^2(R, \mathbb{Z}) \cong \mathbb{Z}$. In order to get this c, we take (i) a simplicial decomposition of R and (ii) its dual cellular decomposition. (We enlarge the cells a bit to get the U_j's and an open covering $R = \bigcup_j U_j$.)

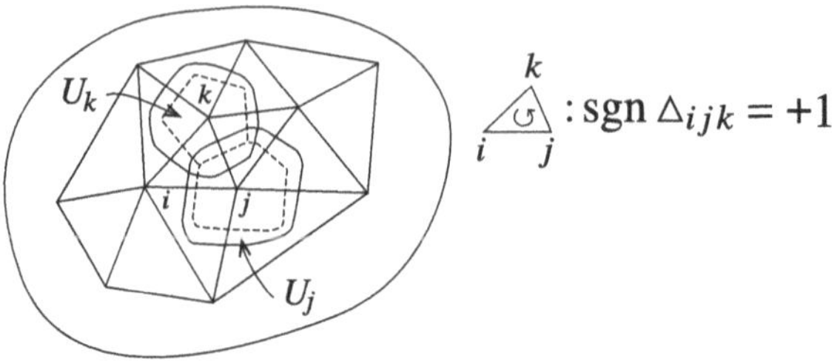

Then we have $c = \sum \operatorname{sgn} \Delta_{ijk} \cdot c_{ijk}$ (the sum is taken over all simplexes Δ_{ijk}). Moreover, we have the following:

3.3.1 $-\int_R \xi = c.$

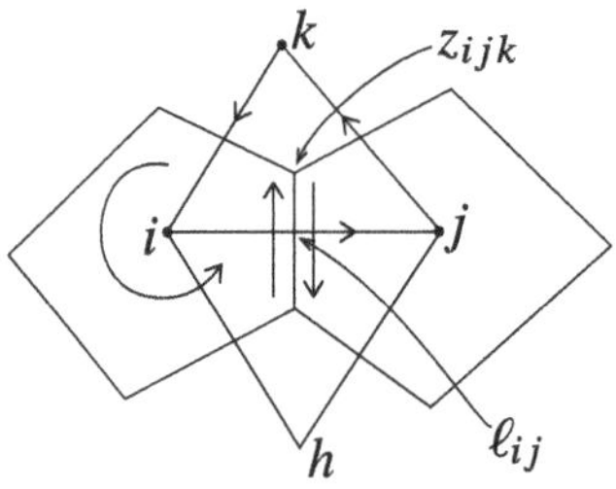

Proof We have

$$
\begin{aligned}
\int_R \xi &= \sum_i \int_{\Box_i} \xi = \sum_i \int_{\Box_i} \mathrm{d}\sigma_i \\
&= \sum_i \oint_{\partial \Box_i} \sigma_i = \sum_{i,j} \int_{\ell_{ij}} (\sigma_i - \sigma_j) && \text{(by Green's formula)} \\
&= -\sum_{i,j} \int_{\ell_{ij}} \mathrm{d}\mu_{ij} = -\sum_{i,j} [\mu_{ij}(z_{ijk}) - \mu_{ij}(z_{ijh})] \\
&= -\sum_{z_{ijk}} [\mu_{ij}(z_{ijk}) + \mu_{jk}(z_{ijk}) + \mu_{ki}(z_{ijk})] \\
&= -\sum_{z_{ijk}} c_{ijk} && (\text{by } \mu_{ij} + \mu_{jk} + \mu_{ki} = c_{ijk}) \\
&= -c,
\end{aligned}
$$

where $\Box_i \subset U_i$ denotes a cell. □

In order to prove $c([\mathfrak{d}]) = \deg \mathfrak{d}$ ($\deg \mathfrak{d} = \sum m_\nu$), it suffices to show that $c([\mathfrak{p}]) = 1$ for any $\mathfrak{p} \in R$.

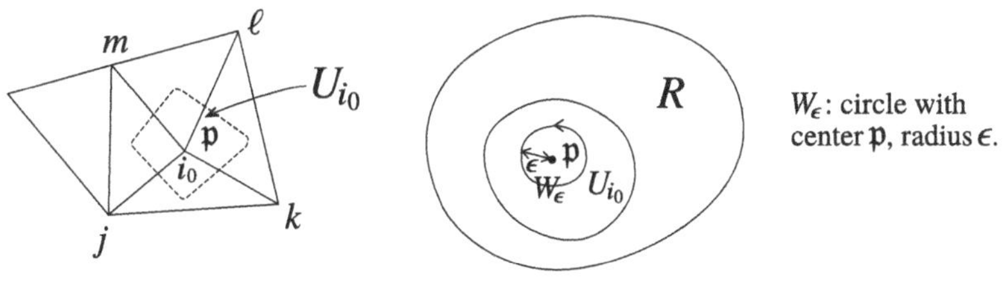

(a) (b)

We take a sufficiently fine open covering $R = U_{i_0} \cup U_j \cup U_k \cup \cdots$ $(\mathfrak{p} \in U_{i_0})$ and let $f_j(z) = 0$ be the equation in U_j defining $\mathfrak{d} = \mathfrak{p}$. That is, for $j \neq i_0$, $f_j(z) = 1$ and $f_{i_0}(z)$ has a simple zero at $\mathfrak{p}$ and $f_{i_0}(z) \neq 0$ for $z \neq \mathfrak{p}$. If we put $f_{jk} = \frac{f_j}{f_k}$, then $[\mathfrak{p}]$ is determined by $\{f_{jk}\}$. We put $c([\mathfrak{p}]) = \{c_{ijk}\}$. Then, from the arguments above, $c_{ijk} = \frac{1}{2\pi \mathrm{i}}(\log f_{ij} + \log f_{jk} + \log f_{ki})$, there is a C^∞ function μ_{ik} on $U_i \cap U_k$ and a C^∞ 1-form σ_i on U_i satisfying $c_{ijk} = \mu_{ij} + \mu_{jk} + \mu_{ki}$ (on $U_i \cap U_j \cap U_k$), $\mathrm{d}\mu_{ij} = \sigma_j - \sigma_i$ (on $U_i \cap U_j$) and $\xi = \mathrm{d}\sigma_i$ (on U_i). Recall that $c([\mathfrak{p}]) = -\int_R \xi$. If we put $\gamma_{ij} = \mu_{ij} - \frac{1}{2\pi \mathrm{i}} \log f_{ij}$, then $\gamma_{ij} + \gamma_{jk} + \gamma_{ki} = 0$ and there exists a C^∞ function τ_i on U_i satisfying $\gamma_{ij} = \tau_j - \tau_i$ on $U_i \cap U_j$. Hence, $\omega = \sigma_i + \frac{1}{2\pi \mathrm{i}}\mathrm{d}\log f_i - \mathrm{d}\tau_i = \sigma_j + \frac{1}{2\pi \mathrm{i}}\mathrm{d}\log f_j - \mathrm{d}\tau_j$ is a 1-form on R. It is clear that ω is C^∞ on $R \setminus \{\mathfrak{p}\}$. We have $\xi = \mathrm{d}\sigma_i = \mathrm{d}\omega$. In view of Fig. b, we have $\displaystyle\int_R \xi = \int_{R\setminus\{\mathfrak{p}\}} \xi = \lim_{\epsilon\to 0}\int_{R\setminus W_\epsilon} \mathrm{d}\omega = -\lim_{\epsilon\to 0}\oint_{\partial W_\epsilon} \frac{1}{2\pi \mathrm{i}}\mathrm{d}\log f_{i_0} = -1$. Hence $c([\mathfrak{p}]) = -\int_R \xi = 1$. □

Proposition 3.4 *Let D be a divisor on S. Then $c([D])$ and D are dual to each other (note that a divisor D is a 2-cycle on S). In other words, for all $a \in H^2(S, \mathbb{Z})$,*

$$c([D]) \cdot a = a(D).$$

Proof By linearity, it suffices to show the assertion for an irreducible curve C instead of a given D.

We take a suitable open covering $S = \bigcup_j U_j$ and let $f_j(z) = 0$ be the minimal local equation of C in U_j. If we put $f_{jk} = f_j/f_k$ on $U_j \cap U_k$, then $[C] = \{f_{jk}\}$. If $c([C]) = \{c_{ijk}\}$, then $c_{ijk} = \frac{1}{2\pi \mathrm{i}}(\log f_{ij} + \log f_{jk} + \log f_{ki})$ and, by de Rham's theorem, we can express c_{ijk} with a 2-form ξ. We go along the same line as in the proof of 3.3 employing similar notation. Putting $\xi = \mathrm{d}\sigma_i = \mathrm{d}\sigma_j = \cdots$, $\omega = \sigma_i + \frac{1}{2\pi \mathrm{i}}\mathrm{d}\log f_i - \mathrm{d}\tau_i$, we obtain $\xi = \mathrm{d}\omega$, where σ_i is a C^∞ 1-form and τ_i is a C^∞ function on U_i. In this way, we get the correspondence $c([D]) \leftarrow \xi$ (not unique but the class of ξ modulo exact forms is unique). Similarly, there is a 2-form η with $d\eta = 0$ such that $a \leftarrow \eta$. By de Rham's theorem, we have $c([C]) \cdot a \leftarrow \xi \wedge \eta$ and $c([C]) \cdot a = \int_S \xi \wedge \eta$, $a(C) = -\int_C a$. Therefore, in order to prove $c([C]) \cdot a = a(C)$, it suffices to show that $\int_S \xi \wedge \eta = -\int_C \eta$.

We can choose η in such a way that η vanishes at any singular points of C (in fact, take a sufficiently small neighborhood W of a singular point and subtract a suitable exact form with support in W from η). ω is C^∞ on $S \setminus C$. Hence

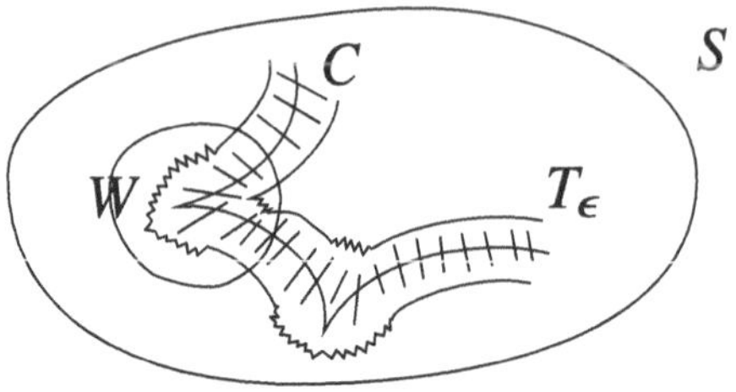

$T_\epsilon \supset C$: tubular neighborhood of C

$$\int_S \xi \wedge \eta = \lim_{\epsilon \to 0} \int_{S \setminus T_\epsilon} \xi \wedge \eta = \lim_{\epsilon \to 0} \int_{S \setminus T_\epsilon} \mathrm{d}(\omega \wedge \eta)$$

$$= - \lim_{\epsilon \to 0} \int_{\partial T_\epsilon} \omega \wedge \eta \overset{(*)}{=} - \lim_{\epsilon \to 0} \int \frac{1}{2\pi \mathrm{i}} \frac{\mathrm{d} f_i}{f_i} \wedge \eta \overset{(**)}{=} - \int_C \eta$$

Here the equality $(*)$ can be obtained by substituting the expression of ω given above. To see that the equality $(**)$ holds, we calculate as follows.

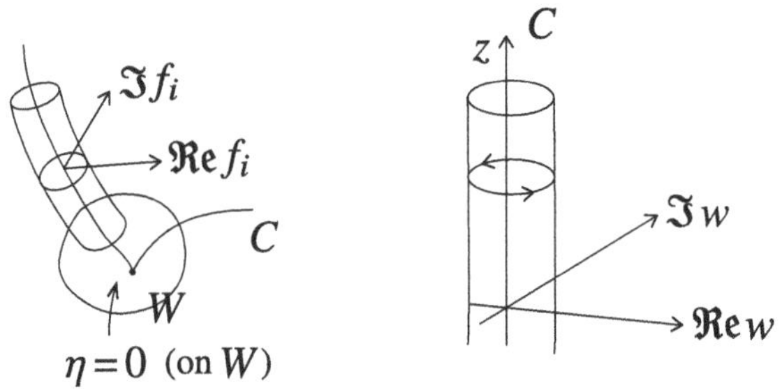

Since $\eta = 0$ at a singular point of C, we consider around a smooth point of C. We let (w, z) denote a system of local coordinates of S around that point. We can assume that $w = f_i$, since $C\colon f_i(z) = 0$. Writing

$$\eta = A\mathrm{d}w \wedge \mathrm{d}z + B\mathrm{d}\bar{w} \wedge \mathrm{d}\bar{z} + C\mathrm{d}w \wedge \mathrm{d}\bar{z} + D\mathrm{d}z \wedge \mathrm{d}\bar{z} + E\mathrm{d}\bar{w} \wedge \mathrm{d}z + F\mathrm{d}w \wedge \mathrm{d}\bar{w},$$

we get

$$\frac{\mathrm{d}w}{w} \wedge \eta = \frac{\mathrm{d}w}{w} \wedge D\mathrm{d}z \wedge \mathrm{d}\bar{z} + (\sharp)\mathrm{d}w \wedge \mathrm{d}\bar{w}.$$

However, fixing a small $\varepsilon > 0$ and putting $w = \varepsilon e^{i\theta}$, we get $dw = \varepsilon i e^{i\theta} d\theta$, $d\bar{w} = -\varepsilon i e^{-i\theta} d\theta$ and, hence, $dw \wedge d\bar{w} = 0$. So, we have

$$\int \frac{df_i}{f_i} \wedge \eta = 2\pi i \int_C D dz \wedge d\bar{z} = 2\pi i \int_C \eta.$$

This completes the proof. □

3.5 Let C and F be an irreducible curve and a complex line bundle over S, respectively, and let $\mu \colon \widetilde{C} \to C \subset S$ be the desingularization of C. We let $\widetilde{F} = \mu^* F$ be the line bundle induced on $\widetilde{C}$. For two line bundles F, G over S, we have $c(F) \cdot c(G) = c(G) \cdot c(F)$.

Proposition 3.5.1 $F \cdot C = c(\widetilde{F})$.

Proof We use 3.4. We have $F \cdot C = c(F) \cdot c([C]) = c([C]) \cdot c(F) = c(F)(C) = c(\widetilde{F})$. □

3.6 Let B, C be irreducible curves on S, $B \neq C$. Then $C \cap B$ consists of a finite number of points. We define the *intersection multiplicity* $I_{\mathfrak{p}}(B, C)$ at $\mathfrak{p} \in C \cap B$.

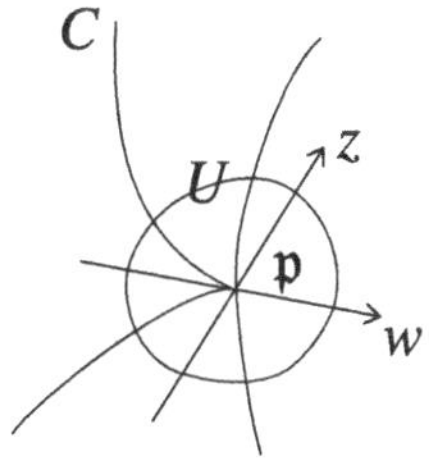

We denote by $R(w, z) = 0$ the minimal local equation of B at $\mathfrak{p}$, $\mu \colon \widetilde{C} \to C$ the desingularization and $\mu^{-1}(\mathfrak{p}) = \{\mathfrak{p}_1, \ldots, \mathfrak{p}_\lambda, \ldots, \mathfrak{p}_r\}$. In a neighborhood of $\mathfrak{p}_\lambda$, we may assume $\mu \colon t_\lambda \to (w, z) = (P_\lambda(t_\lambda), t_\lambda^{m_\lambda})$ (cf. 1.2, take a suitable t_λ). Put $R(P_\lambda(t_\lambda), t_\lambda^{m_\lambda}) = t_\lambda^{n_\lambda}(a_{\lambda 0} + a_{\lambda 1} t_\lambda + \cdots)$ $(a_{\lambda 0} \neq 0)$. Obviously, n_λ is determined by $\mathfrak{p}_\lambda$.

Definition 3.6.1 $I_{\mathfrak{p}}(B, C) = \sum_{\mathfrak{p}_\lambda \in \mu^{-1}(\mathfrak{p})} n_\lambda$.

Here we remark that $I_{\mathfrak{p}}(B, C) = I_{\mathfrak{p}}(C, B)$.

Proposition 3.7 *For irreducible curves B, C on S,*

$$B \cdot C = \sum_{\mathfrak{p} \in B \cap C} I_{\mathfrak{p}}(B, C).$$

Proof Taking a suitable open covering $S = \bigcup U_j$, we let $F = [B]$ be defined by $\{f_{jk}\} = \left\{\frac{R_j}{R_k}\right\}$. We take an open covering $\widetilde{C} = \bigcup \widetilde{U}_\lambda$ in such a way that

$\mu(\widetilde{U}_\lambda) \subset U_j = U_{j(\lambda)}$. Put $\widetilde{R}_\lambda(t) = R_{j(\lambda)}(\mu(t))$ on $\widetilde{U}_\lambda$ and $\tilde{f}_{\lambda\nu}(t) = f_{j(\lambda)j(\nu)}(\mu(t))$ on $\widetilde{U}_\lambda \cap \widetilde{U}_\nu$. Hence, putting $\tilde{f}_{\lambda\nu}(t) = \frac{\widetilde{R}_\lambda(t)}{\widetilde{R}_\nu(t)}$, $\widetilde{F}$ is defined by $\{\tilde{f}_{\lambda\nu}\}$. $\tilde{\varphi} = \{\widetilde{R}_\lambda\}$ is a holomorphic section of $\widetilde{F}$ and, putting $\mathfrak{d} = (\tilde{\varphi})$, we have $\widetilde{F} = [\mathfrak{d}]$. Therefore, from 3.5.1 and 3.3, we have $B \cdot C = F \cdot C = c(\widetilde{F}) = \deg \mathfrak{d}$. On the other hand, $\deg \mathfrak{d} = \sum n_\lambda = \sum_{\mathfrak{p} \in B \cap C} I_{\mathfrak{p}}(B, C)$ (because, taking $\widetilde{U}_\lambda$ as a coordinate neighborhood around $\mathfrak{p}_\lambda$, we have $\widetilde{R}_\lambda(t) = R(P_\lambda(t_\lambda), t_\lambda^{m_\lambda}) = t_\lambda^{n_\lambda}(a_{\lambda 0} + a_{\lambda 1} t_\lambda + \cdots)$, $a_{\lambda 0} \neq 0$). □

3.8 We give the definition of the canonical bundle K over S. Let $(z_j^1, z_j^2) = (w_j, z_j)$ be a system of local coordinates on a coordinate neighborhood U_j, $S = \bigcup U_j$. $\varphi = \{\varphi_j(z) dz_j^1 \wedge dz_j^2\}$ is said to be a meromorphic differential form if:

(a) $\varphi_j(z)$ is a meromorphic function on U_j, and
(b) $\varphi_j dz_j^1 \wedge dz_j^2 = \varphi_k dz_k^1 \wedge dz_k^2$.

Definition 3.8.1 $D = (\varphi) = (\varphi_j)$ is called a *canonical divisor*. It is uniquely determined modulo linear equivalence.

Definition 3.8.2 $K = [D]$ is called the *canonical bundle* over S.

If we put $J_{jk}(z) = \det \frac{\partial(z_k^1, z_k^2)}{\partial(z_j^1, z_j^2)}$, then $\varphi_j = J_{jk}\varphi_k$ and we see that φ is a meromorphic section of the line bundle defined by the 1-cocycle $\{J_{jk}\}$. Hence one may say that K is defined by $\{J_{jk}\}$.

Definition 3.8.3 $c_1 = -c(K)$ is called the first Chern class of S.

Example 3.8.4 In the situation of 3.5, $\widetilde{C}$ is a Riemann surface. We let t_λ be a local uniformizing parameter on $\widetilde{U}_\lambda$, $\widetilde{C} = \bigcup_{\lambda \in \Lambda} \widetilde{U}_\lambda$. The canonical bundle $\mathfrak{k}$ over $\widetilde{C}$ is determined by the 1-cocyle $\{\mathfrak{k}_{\lambda\nu}\} = \left\{\frac{dt_\nu}{dt_\lambda}\right\}$. If we denote the genus of $\widetilde{C}$ by $\pi(\widetilde{C})$, then $c(\mathfrak{k}) = 2\pi(\widetilde{C}) - 2$ from the Riemann–Roch theorem.

Adjunction formula 3.9 *Let C be an irreducible curve, $\mu \colon \widetilde{C} \to C \subset S$ the desingularization, K and $\mathfrak{k}$ the canonical bundles over S and $\widetilde{C}$, respectively, and let $\mathfrak{c}$ denote the conductor of C. Then,*

$$\mathfrak{k} = \mu^*[C] + \mu^* K - [\mathfrak{c}].$$

Proof We take a sufficiently fine open covering $S = \bigcup_j U_j$, $\widetilde{C} = \bigcup_\lambda \widetilde{U}_\lambda$ and system of local coordinates (w_j, z_j), t_λ on U_j, $\widetilde{U}_\lambda$, respectively, so that the following (a), (b), and (c) hold:

(a) $\mu(\widetilde{U}_\lambda) \subset U_j$ $(j = j(\lambda))$.
(b) If $\mu(\widetilde{U}_\lambda) \cap \mathfrak{s} = \emptyset$, then $U_{j(\lambda)} \cap \mathfrak{s} = \emptyset$, where $\mathfrak{s} = \mathrm{Sing}(C)$.
In this case, we may put $\mu \colon t_\lambda \to (w_j, z_j) = (0, t_\lambda)$ on $\widetilde{U}_\lambda$. Hence the minimal local equation of C on U_j is $R_j(w_j, z_j) = w_j = 0$.

(c) If $\mu(\widetilde{U}_\lambda) \ni \mathfrak{p}$ ($\mathfrak{p} \in \mathfrak{s}$), then $\mathfrak{p}_\lambda \colon t_\lambda = 0$, $\mathfrak{p} = \mu(0) = (0, 0)$ and μ can be expressed on $\widetilde{U}_\lambda$ as

$$\mu \colon t_\lambda \to (w_j, z_j) = (P_\lambda(t_\lambda), t_\lambda^{m_\lambda}).$$

We denote the minimal local equation of C on U_j by $R_j(w_j, z_j) = 0$.

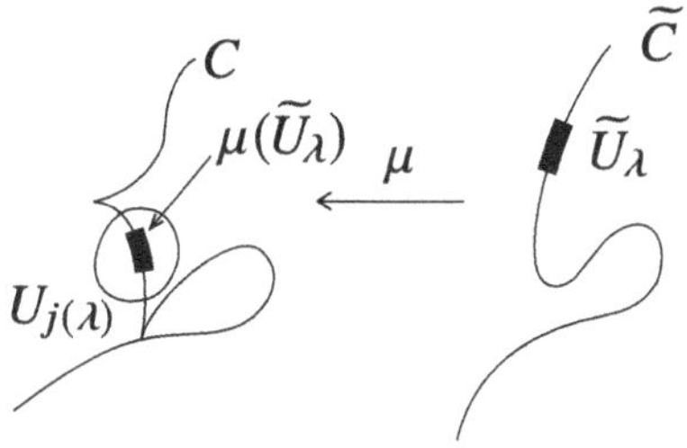

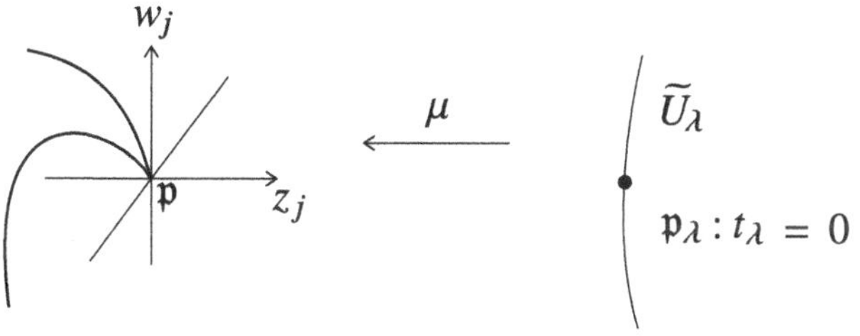

In the notation of 1.3, we have $\sigma_\lambda \mathrm{d}t_\lambda = \dfrac{\mathrm{d}(t^{m_\lambda})}{\partial_{w_j} R_j(P_\lambda(t_\lambda), t_\lambda^{m_\lambda})}$ (where $\partial_{w_j} R_j = \frac{\partial R_j}{\partial w_j}$) and $\sigma_\lambda = \frac{1}{t_\lambda^{c_\lambda}}(a_0 + a_1 t_\lambda + \cdots)$ $(a_0 \neq 0)$. In particular, on $\widetilde{U}_\lambda$ with $\mu(\widetilde{U}_\lambda) \cap \mathfrak{s} = \emptyset$, we have $\partial w_j R_j = \frac{\partial w_j}{\partial w_j} = 1$ $(\mu(\widetilde{U}_\lambda) \subset U_j)$. Hence, since $\mathfrak{c} = \sum_\nu c_\nu \mathfrak{p}_\nu$, we see that $\sigma^{-1} = 0$ is a local equation of $\mathfrak{c}$. It follows that $[\mathfrak{c}]$ is determined by the 1-cocycle $\left\{\frac{\sigma_\lambda^{-1}}{\sigma_\nu^{-1}}\right\}$. In view of 3.7 and 3.8.1, we have $[C] = \{f_{jk}\}$, $K = \{J_{jk}\}$, $\mu^*[C] = \{\tilde{f}_{\lambda\nu}\}$ and $\mu^* K = \{\widetilde{J}_{\lambda\nu}\}$, where $f_{jk} = \frac{R_j}{R_k}$, $\tilde{f}_{\lambda\nu}(t) = f_{j(\lambda)j(\nu)}(\mu(t))$ and $\widetilde{J}_{\lambda\nu} = J_{j(\lambda)j(\nu)}(\mu(t))$. We define $g_{jk}(z)$ by the following (d):

(d) $\mathrm{d}R_j \wedge \mathrm{d}z_j = g_{jk}(z)\mathrm{d}R_k \wedge \mathrm{d}z_k$.

Then, since $\mathrm{d}R_j = \partial_{w_j} R_j \mathrm{d}w_j + \partial_{z_j} R_j \mathrm{d}z_j$, substituting it for (d), we get $\partial_{w_j} R_j \mathrm{d}w_j \wedge \mathrm{d}z_j = g_{jk} \partial_{w_k} R_k \mathrm{d}w_k \wedge \mathrm{d}z_k$. Hence,

(e) $\partial_{w_j} R_j = g_{jk} J_{jk} \partial_{w_k} R_k$.

On the other hand, since $R_j = f_{jk} R_k$, we have $\mathrm{d}R_j(z) = f_{jk}(z)\mathrm{d}R_k(z) + R_k(z)\mathrm{d}f_{jk}$. Hence, for $z \in C$, $\mathrm{d}R_j(z) = f_{jk}(z)\mathrm{d}R_k(z)$. This and (d) imply that $f_{jk}\mathrm{d}z_j = g_{jk}\mathrm{d}z_k$ for $z \in C$. Applying μ^*, we get

(f) $\tilde{f}_{\lambda\nu}(t)\mathrm{d}(t_\lambda^{m_\lambda}) = g_{j(\lambda)j(\nu)}(\mu(t))\mathrm{d}(t_\nu^{m_\nu})$.

By (e) and (f),

$$\tilde{f}_{\lambda\nu}(t)\frac{\mathrm{d}(t_\lambda^{m_\lambda})}{\partial_{w_{j(\lambda)}}R_{j(\lambda)}(P_\lambda(t_\lambda),t_\lambda^{m_\lambda})}=\frac{1}{\widetilde{J}_{\lambda\nu}}\frac{\mathrm{d}(t_\nu^{m_\nu})}{\partial_{w_{j(\nu)}}R_{j(\nu)}(P_\nu(t_\nu),t_\nu^{m_\nu})}$$

and, hence, $\tilde{f}_{\lambda\nu}\sigma_\lambda \mathrm{d}t_\lambda = \frac{1}{\widetilde{J}_{\lambda\nu}}\sigma_\nu \mathrm{d}t_\nu$. It follows that $\mathfrak{k}_{\lambda\nu} = \frac{\mathrm{d}t_\nu}{\mathrm{d}t_\lambda} = \tilde{f}_{\lambda\nu}\cdot\widetilde{J}_{\lambda\nu}\cdot\left(\frac{\sigma_\lambda}{\sigma_\nu}\right)$. Considering the corresponding line bundles, we get

$$\mathfrak{k}=\mu^*[C]+\mu^*K-[\mathfrak{c}]$$

as wished. □

Corollary 3.10 *Under the notation in 3.9,*

$$\pi(\widetilde{C})=\frac{1}{2}(C^2+KC)+1-\frac{1}{2}\deg\mathfrak{c}.$$

In particular, if C has no singular points, then

$$\pi(C)=\frac{1}{2}(C^2+KC)+1.$$

Proof We have $c(\mu^*F) = FC$, $c(\mu^*[C]) = C\cdot C = C^2$, $c([\mathfrak{d}]) = \deg\mathfrak{d}$ by 3.3, 3.5.1. By the Riemann–Roch theorem, $c(\mathfrak{k}) = 2\pi(\widetilde{C})-2$, while we have $c(\mathfrak{k}) = C^2+KC-\deg\mathfrak{c}$ by 3.9. □

Definition 3.11 For a divisor D and a line bundle F over S,

$$\pi(D)=\frac{1}{2}(D^2+KD)+1,$$

$$\pi(F)=\frac{1}{2}(F^2+KF)+1.$$

We call them the *virtual genera* of D, F, respectively.

Proposition 3.12 *For any irreducible curve C,*

$$\pi(C)=\pi(\widetilde{C})+\frac{1}{2}\deg\mathfrak{c}.$$

In particular, the following (a), (b) *and* (c) *hold:*

(a) $\pi(C)>\pi(\widetilde{C})\Longleftrightarrow C\neq\widetilde{C}$.
(b) $\pi(C)=\pi(\widetilde{C})\Longleftrightarrow C=\widetilde{C}$.
(c) $\pi(C)=0\Longleftrightarrow C$: *a non-singular rational curve.*

1.4 Riemann–Roch Theorem

Notation 4.0 For an algebraic (analytic) sheaf Ξ, a line bundle F and a divisor D, we employ the following notation:

$$\chi(S, \Xi) = \sum_{\nu}(-1)^{\nu} \dim H^{\nu}(S, \Xi),$$

$$\chi(S, F) = \chi(S, \mathcal{O}(F)),$$

$$\chi(S, D) = \chi(S, \mathcal{O}([D])).$$

Definition 4.1

(a) $q = \dim H^1(S, \mathcal{O})$: the *irregularity* of S. It holds that $q = \frac{1}{2}b_1$ (b_1 : the first Betti number of S).
(b) $p_g = \dim H^2(S, \mathcal{O})$: the *geometric genus* of S.
(c) $p_a = p_g - q$: the *arithmetic genus* of S.

Remark 4.1.1 On the naming of "irregularity"(?): Let M be a surface defined by $f(z_0, z_1, z_2, z_3) = 0$ in $\mathbb{P}^3$, and $S = \widetilde{M}$ its non-singular model. Then we have $q(S) = p_g - p_a = 0$ in almost all cases. In fact, for $M\colon f(z_0, \ldots, z_3) = 0$, if we consider the correspondence $f \longleftrightarrow (a_1, a_2, \ldots, a_N) \in \mathbb{C}^N$ (the coefficients of f), then the measure of $\{f \mid b_1(S) > 0\}$ is 0. (Recall $b_1 = 2q$.)

Riemann–Roch Theorem 4.2

$$\chi(S, D) = \frac{1}{2}(D^2 - KD) + \chi(S, \mathcal{O}),$$

i.e., $\chi(S, D) = \frac{1}{2}(D^2 - KD) + p_a + 1$. (*The latter follows from* $\dim H^0(S, \mathcal{O}) = 1$ *and 4.1.)*

Proof Put $\psi(F) = \chi(S, F) - \frac{1}{2}(F^2 - KF)$. If we have $\psi(F - [C]) = \psi(F)$ for any irreducible curve C, then, since we can express as $D = \sum m_i C_i$, we will have $\psi([D]) = \psi(0) = \chi(S, \mathcal{O}) = p_a + 1$ and complete the proof. Therefore, it suffices to prove the following lemma. □

Lemma 4.2.1 $\psi(F - [C]) = \psi(F)$.

Proof We have the following exact sequences (1.14):

(a) $0 \to \mathcal{O}(F - [C]) \to \mathcal{O}(F) \to \mathcal{O}(F)_C \to 0,$
(b) $0 \to \mu_*\mathcal{O}(\mu^* F - [\mathfrak{c}]) \to \mathcal{O}(F)_C \to \mathcal{M} \to 0.$

From (a), we have $\chi(S, F) - \chi(S, F - [C]) = \chi(C, \mathcal{O}(F)_C)$. From (b), we have $\chi(C, \mathcal{O}(F)_C) = \chi(\widetilde{C}, \mu^* F - [\mathfrak{c}]) + \dim H^0(C, \mathcal{M})$. We infer from 1.15 and the Riemann–Roch theorem for curves that $\dim H^0(C, \mathcal{M}) = \frac{1}{2}\deg \mathfrak{c}$ and $\chi(\widetilde{C}, \mu^* F - [\mathfrak{c}]) = c(\mu^* F - [\mathfrak{c}]) - \pi(\widetilde{C}) + 1 = F \cdot C - \deg \mathfrak{c} - \pi(\widetilde{C}) + 1$, respectively. Therefore, using 3.12, we get $\chi(S, F) - \chi(S, F - [C]) = F \cdot C - \pi(C) + 1$. On the other hand, $\frac{1}{2}(F^2 - KF) - \frac{1}{2}((F - C)^2 - K(F - C)) = F \cdot C - \frac{1}{2}(C^2 + K \cdot C) = F \cdot C - \pi(C) + 1$ (3.11). Hence, by the definition of $\psi(\cdot)$, we get $\psi(F) - \psi(F - [C]) = 0$. □

Riemann–Roch–Hirzebruch Formula 4.3

$$\chi(S, D) = \frac{1}{2}(D^2 - KD) + \frac{1}{12}(c_1^2 + c_2).$$

(c_ν: *the* ν*-th Chern class of* S. c_2 *equals the Euler number of* S.)

By virtue of 4.2, it is sufficient for this to prove the following formula:

Noether's formula 4.4 $p_a + 1 = \frac{1}{12}(c_1^2 + c_2)$.

Proof (Enriques [2]) It suffices to show Theorem 4.6 below. We will prepare things in 4.5 for that. □

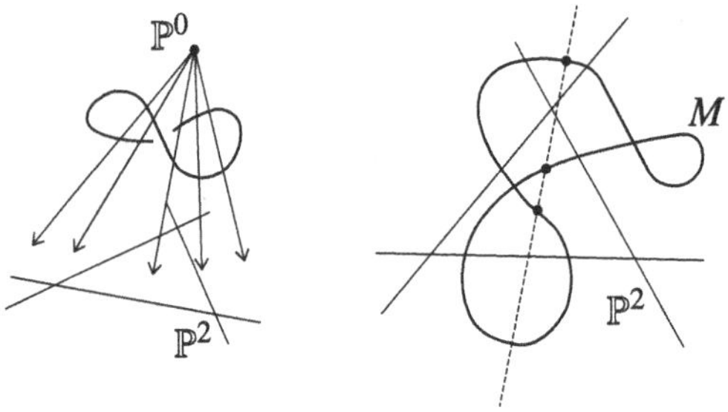

4.5 On generic projections.

(I) Projection of a non-singular curve $C \subset \mathbb{P}^3$.

Take linear subspaces $\mathbb{P}^0$, $\mathbb{P}^2 \subset \mathbb{P}^3$ in general position. We let M be the image of C under the projection $\lambda\colon \mathbb{P}^3 \to \mathbb{P}^2$ with center $\mathbb{P}^0$. Then the singular points of M are at most nodes; we let d denote the number of nodes. We call the degree of the irreducible equation defining M the degree of M, and put it n. In fact, n equals the number of intersection points of M and a general line in $\mathbb{P}^2$ (that is, the number of generic hyperplane section of M). Then $\pi(C) = \frac{1}{2}(n-1)(n-2) - d$ as is well-known.

(II) Projection of a non-singular surface $S \hookrightarrow \mathbb{P}^N$.

We take linear subspaces $\mathbb{P}^3$, $\mathbb{P}^{N-4} \hookrightarrow \mathbb{P}^N$ in general position. We let M be the image of S under the projection $\lambda\colon \mathbb{P}^N \to \mathbb{P}^3$ with center $\mathbb{P}^{N-4}$. Then M is a model of S. We study $\Delta = \mathrm{Sing}(M)$.

If we take a system of local coordinates (x, y, z) of $\mathbb{P}^3$ around $\mathfrak{p} \in \Delta$ suitably, then the equation of M in a neighborhood of $\mathfrak{p}$ can take one of the following three forms:

$$
\begin{array}{ll}
\text{(a) } yz = 0 & (\mathfrak{p}\colon \text{an ordinary double point}),\\
\text{(b) } xyz = 0 & (\mathfrak{p}\colon \text{a triple point}),\\
\text{(c) } xy^2 - z^2 = 0 & (\mathfrak{p}\colon \text{a cuspidal (or pinch) point}).
\end{array}
$$

In the following, we study these three cases separately. We employ notation such as $\lambda^{-1}(\Delta) = \widetilde{\Delta}$, $\lambda^{-1}(\mathfrak{p}) = \{\tilde{\mathfrak{p}}_1, \dots\}$.

(a) Around a double point, taking a suitable system of local coordinates (u, v) on S, λ can be expressed as

$$\lambda\colon (u, v) \longrightarrow (x, y, z) = (u, 0, v) \quad \text{in a neighborhood of } \tilde{\mathfrak{p}}_1,$$

$$\lambda\colon (u, v) \longrightarrow (x, y, z) = (u, v, 0) \quad \text{in a neighborhood of } \tilde{\mathfrak{p}}_2.$$

(b) Around a triple point, similarly to (a), we have

$$\lambda\colon (u, v) \longrightarrow (x, y, z) = (u, v, 0) \quad \text{in a neighborhood of } \tilde{\mathfrak{p}}_1,$$

$$\lambda\colon (u, v) \longrightarrow (x, y, z) = (u, 0, v) \quad \text{in a neighborhood of } \tilde{\mathfrak{p}}_2,$$

$$\lambda\colon (u, v) \longrightarrow (x, y, z) = (0, u, v) \quad \text{in a neighborhood of } \tilde{\mathfrak{p}}_3.$$

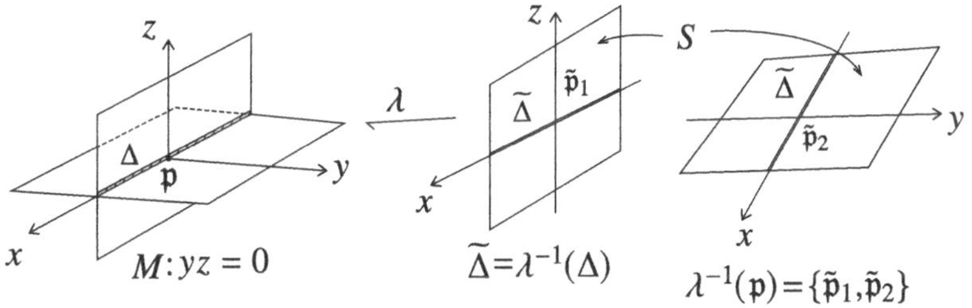

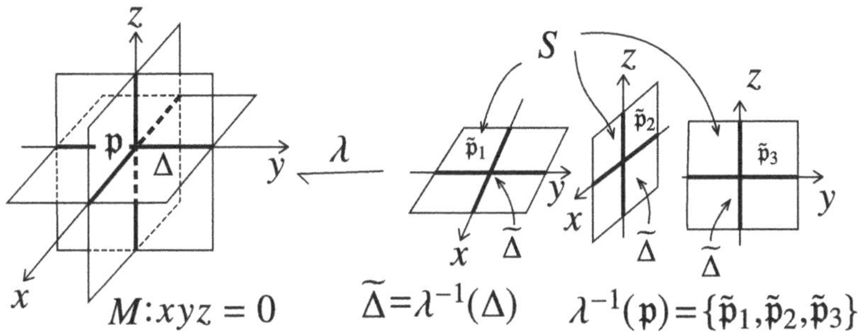

(c) Around a cuspidal point, taking a suitable system of local coordinates (u, v) on S, λ can be expressed as

$$\lambda\colon (u, v) \longrightarrow (x, y, z) = (u^2, v, uv)$$

in a neighborhood of $\tilde{\mathfrak{p}}$.

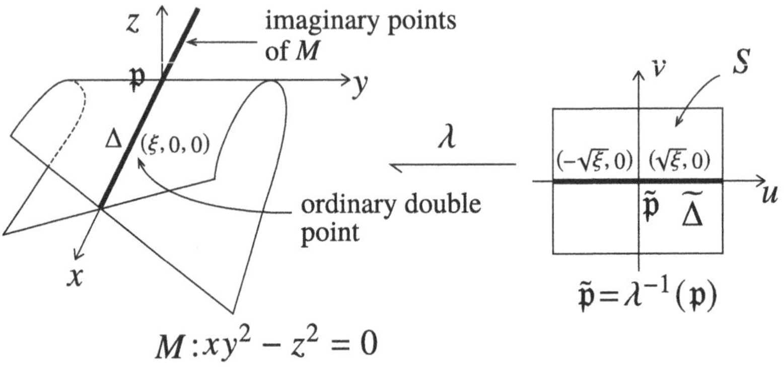

As is easily seen, the cuspidal point of M is a simple point of Δ.

We put:

- n: the degree of M (that is, the degree of the irreducible equation defining M).

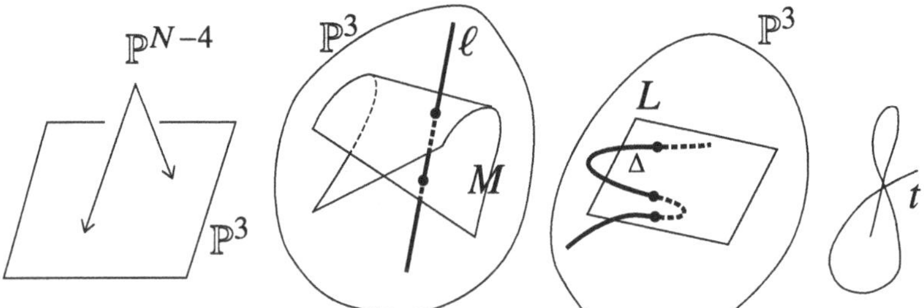

- m: the degree of Δ (Δ is the double curve).
  In fact, if we denote by ℓ and L a general line and a general plane, respectively, then

$$n = \mathrm{Card}(M \cap \ell), \quad m = \mathrm{Card}(\Delta \cap L).$$

- t: the number of triple points of M. (From the above observations (a), (b), and (c), any singular point of Δ is a triple point of M.)
- $\Delta = \bigcup_{\nu=1}^{\rho} \Delta_\nu$: the decomposition of Δ into irreducible curves Δ_ν. We denote by $\hat{\Delta}_\nu$ the non-singular model of Δ_ν.

- Then $\pi(\Delta) = \sum_\nu \{\pi(\hat{\Delta}_\nu) - 1\} + 1 + 2t$.
  (If $\mathcal{O}$ denotes the sheaf of germs of holomorphic functions on $\mathbb{P}^3$ and if $\mathcal{O}_\Delta$ is the restriction of $\mathcal{O}$ to Δ, then $\chi(\Delta, \mathcal{O}_\Delta) = \dim H^0(\Delta, \mathcal{O}_\Delta) - \dim H^1(\Delta, \mathcal{O}_\Delta) = 1 - \pi(\Delta)$.)

Theorem 4.6 (Enriques [2]) *Let the notation be as in* 4.5, (II). *Then,*

$$p_a = \binom{n-1}{3} - (n-4)m + \pi(\Delta) - 1,$$

$$c_1^2 = n(n-4)^2 - (5n-24)m + 4\pi(\Delta) - 4 + t,$$

$$c_2 = n(n^2 - 4n + 6) - (7n-24)m + 8\pi(\Delta) - 8 - t,$$

where p_a: *the arithmetic genus of* S, *and* c_1, c_2: *Chern classes of* S.

Proof We will calculate p_a in 4.12. We devote 4.8 and 4.9 to the preparations. The calculation of c_1^2 and c_2 will be done in 4.13. □

Corollary 4.7 $c_1^2 + c_2 = 12(p_a + 1)$.

Adjunction formula 4.8 *Let* $S \xrightarrow{\lambda} M \hookrightarrow \mathbb{P}^3$ *be the generic projection. If* K *and* $\mathbb{K}$ *respectively denote the canonical bundles over* S *and* $\mathbb{P}^3$, *then*

$$K = \lambda^*\mathbb{K} + \lambda^*[M] - [\widetilde{\Delta}],$$

where $\Delta = \mathrm{Sing}(M)$ *and* $\widetilde{\Delta} = \lambda^{-1}(\Delta)$ ($[\widetilde{\Delta}]$: *the conductor*).

Proof Similar to the case of one dimension lower (3.9). If we put locally

$$\lambda\colon (u, v) \longrightarrow (x, y, z) = (\mu_1(u, v), \mu_2(u, v), \mu_3(u, v)),$$

$$R(x, y, z) = 0\colon \text{the minimal local equation of } M,$$

$$\sigma = \frac{d\mu_1(u, v) \wedge d\mu_2(u, v)}{\frac{\partial R}{\partial z}(\mu_1(u, v), \mu_2(u, v), \mu_3(u, v))},$$

then the conductor is given by $[\widetilde{\Delta}] = -(\sigma)$. For example, around a cuspidal point, we have

$$R = xy^2 - z^2 = 0,$$

$$\lambda\colon (u, v) \longrightarrow (x, y, z) = (u^2, v, uv),$$

$$\sigma = \frac{d(u^2) \wedge dv}{-2z} = -\frac{1}{2}\frac{2u du \wedge dv}{uv} = -\frac{1}{v} du \wedge dv.$$

Therefore, $[\widetilde{\Delta}] = -(\sigma) = (v) = -\left(\frac{1}{v}\right)$. □

From now on until 4.13, we let λ, S, M, Δ, n, m, t, ..., etc. be as in 4.5, (II).

4.9 Preparations for calculating p_g and p_a.

Let $L\colon a_0\zeta_0 + a_1\zeta_1 + a_2\zeta_2 + a_3\zeta_3 = 0$ be a plane in $\mathbb{P}^3$ and put $\mathbb{E} = [L]$. Then $M \approx nL$ (linearly equivalent) and $[M] = n\mathbb{E}$. On the other hand, the canonical bundle over $\mathbb{P}^3$ is $\mathbb{K} = -4\mathbb{E}$ (in fact, the canonical bundle over $\mathbb{P}^N$ is $\mathscr{O}_{\mathbb{P}^N}(-N-1)$). Hence, it follows from 4.8 that

4.9.1 $K = (n-4)\lambda^*\mathbb{E} - [\widetilde{\Delta}]$.

For any line bundle $\mathbb{F}$ over $\mathbb{P}^3$, we can define $\mathscr{O}_{\mathbb{P}^3}(\mathbb{F})$, $\mathscr{O}_{\mathbb{P}^3}(\mathbb{F}-\Delta)$, $\mathscr{O}_{\mathbb{P}^3}(\mathbb{F}-M)$ in a similar way to S (1.10). Moreover, we put

$$\mathscr{O}_{\mathbb{P}^3}(\mathbb{F})_\Delta = \mathscr{O}_{\mathbb{P}^3}(\mathbb{F})/\mathscr{O}_{\mathbb{P}^3}(\mathbb{F}-\Delta),$$
$$\mathscr{O}_{\mathbb{P}^3}(\mathbb{F}-\Delta)_M = \mathscr{O}_{\mathbb{P}^3}(\mathbb{F}-\Delta)/\mathscr{O}_{\mathbb{P}^3}(\mathbb{F}-M).$$

Hence we obtain the following exact sequences 4.9.2, 4.9.3:

4.9.2 $0 \to \mathscr{O}_{\mathbb{P}^3}(\mathbb{F}-M) \to \mathscr{O}_{\mathbb{P}^3}(\mathbb{F}-\Delta) \to \mathscr{O}_{\mathbb{P}^3}(\mathbb{F}-\Delta)_M \to 0,$

4.9.3 $0 \to \mathscr{O}_{\mathbb{P}^3}(\mathbb{F}-\Delta) \to \mathscr{O}_{\mathbb{P}^3}(\mathbb{F}) \to \mathscr{O}_{\mathbb{P}^3}(\mathbb{F})_\Delta \to 0.$

We have:

4.9.4 $\lambda_*(\mathscr{O}(\lambda^*\mathbb{F}-\widetilde{\Delta})) \xleftarrow{\simeq} \mathscr{O}_{\mathbb{P}^3}(\mathbb{F}-\Delta)_M$.

Proof The natural λ homorphism $\rho\colon \mathscr{O}_{\mathbb{P}^3}(\mathbb{F}-\Delta) \to \mathscr{O}(\lambda^*\mathbb{F}-\widetilde{\Delta})$ is given by $\rho\colon \varphi(x,y,z) \to (\lambda^*\varphi)(u,v)$. So, we need to show that it is an isomorphism on the stalk over any point on M. For this purpose, it suffices to consider $\mathfrak{p} \in \Delta$. Here, we consider the case that $\mathfrak{p}$ is a cuspidal point (and the other cases are left to the readers). Then $(\lambda^*\varphi)(u,v) = \varphi(u^2, v, uv)$. (We follow Fig. c in 4.5(II).)

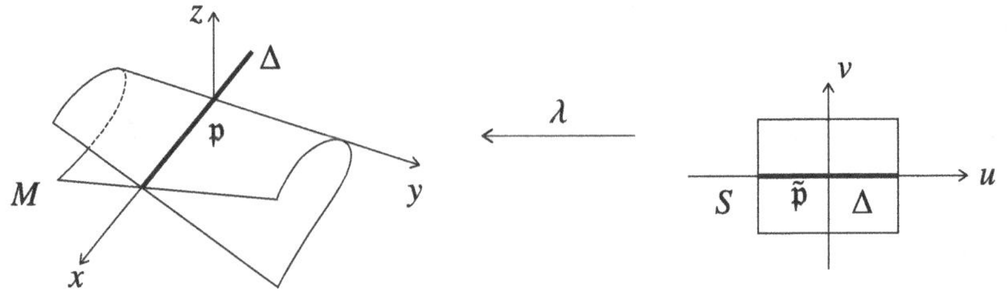

(a) ρ is surjective: Suppose that we are given $\psi(u,v) \in \mathscr{O}(\lambda^*\mathbb{F}-\widetilde{\Delta})_{\tilde{\mathfrak{p}}}$. Since ψ vanishes along $\widetilde{\Delta}$, we have $\psi(u,0) = 0$ (recall that in a neighborhood of $\tilde{\mathfrak{p}}$, $\widetilde{\Delta}\colon v = 0$). Hence we can write $\psi(u,v) = \omega(u,v)v$ with a holomorphic function $\omega(u,v)$. Since $\lambda\colon (u,v) \to (x,y,z) = (u^2, v, uv)$, we can find some $\varphi(x,y,z) \in \mathscr{O}_{\mathbb{P}^3}(\mathbb{F})_{\mathfrak{p}}$ such that $\psi(u,v) = \varphi(u^2,v,uv) = (\lambda^*\varphi)(u,v)$. Moreover, since $\varphi(x,0,0) = \psi(u,0) = 0$ (recall that $\Delta\colon y = z = 0$ in a neighborhood of $\mathfrak{p}$), we have $\varphi(x,y,z) \in (\mathscr{O}_{\mathbb{P}^3}(\mathbb{F}-\Delta)_M)_{\mathfrak{p}}$.

(b) The injectivity of ρ clearly follows from the fact that φ modulo $(xy^2 - z^2)$ can be determined uniquely.

□

Remark 4.9.5 For $\widetilde{C} \xrightarrow{\mu} C \hookrightarrow S$, recall that we have an exact sequence

$$0 \to \mu_*(\mathcal{O}(\mu^* F - \mathfrak{c})) \to \mathcal{O}(F)_C \xrightarrow{\kappa} \mathcal{M} \to 0 \ (1.14)$$

that is, an isomorphism $\mu_*(\mathcal{O}(\mu^* F - \mathfrak{c})) \overset{\cong}{\longleftarrow} \mathrm{Ker}(\mathcal{O}(F)_C \to \mathcal{M})$. 4.9.4 is an extension of it to the one dimension higher case. In fact, $\mathrm{Ker}(\kappa) = \mathcal{O}(F - \mathrm{Sing}(C))_C$.

Notation 4.10 $\mathcal{L}_k(-\Delta) = \left\{ f \;\middle|\; \begin{array}{l} f = f(\zeta) \text{ is a homogeneous polynomial} \\ \text{of degree } k \text{ satisfying } f(\xi) = 0 \text{ for } \xi \in \Delta \end{array} \right\}$.

4.11 $p_g = \dim \mathcal{L}_{n-4}(-\Delta)$.

Proof Put $\mathbb{F} = (n-4)\mathbb{E}$. Then we get

$$\begin{aligned} p_g = \dim H^2(S, \mathcal{O}) &= \dim H^0(S, \mathcal{O}(K)) && \text{(by Serre duality)} \\ &= \dim H^0(S, \mathcal{O}(\lambda^*\mathbb{F} - [\widetilde{\Delta}])) && \text{(by 4.9.1)} \\ &= \dim H^0(M, \mathcal{O}_{\mathbb{P}^3}(\mathbb{F} - \Delta)_M) && \text{(by 4.9.4).} \end{aligned}$$

On the other hand, since $[M] = n\mathbb{E}$, we have $\mathcal{O}_{\mathbb{P}^3}(\mathbb{F} - M) = \mathcal{O}_{\mathbb{P}^3}(-4\mathbb{E})$. It follows from $H^\nu(\mathbb{P}^3, \mathcal{O}_{\mathbb{P}^3}(k\mathbb{E})) = 0$ $(\nu = 1, 2)$ and $H^0(\mathbb{P}^3, \mathcal{O}_{\mathbb{P}^3}(-4\mathbb{E})) = 0$ that $H^0(\mathbb{P}^3, \mathcal{O}_{\mathbb{P}^3}(\mathbb{F} - \Delta)) \overset{\cong}{\to} H^0(M, \mathcal{O}_{\mathbb{P}^3}(\mathbb{F} - \Delta)_M)$. Therefore, $p_g = \dim H^0(\mathbb{P}^3, \mathcal{O}_{\mathbb{P}^3}(\mathbb{F} - \Delta)) = \dim \mathcal{L}_{n-4}(-\Delta)$, because $\mathbb{F} = (n-4)[L]$. □

4.12 $p_a = \binom{n-1}{3} - (n-4)m + \pi(\Delta) - 1$.

Proof We put $\mathbb{F} = (n-4)\mathbb{E}$. Then

$$\begin{aligned} p_a + 1 = \chi(S, \mathcal{O}) &= \chi(S, \mathcal{O}(K)) && \text{(by Serre duality)} \\ &= \chi(S, \mathcal{O}(\lambda^*\mathbb{F} - \widetilde{\Delta})) && \text{(by 4.9.1)} \\ &= \chi(M, \mathcal{O}_{\mathbb{P}^3}(\mathbb{F} - \Delta)_M) && \text{(by 4.9.4)} \\ &= \chi(\mathbb{P}^3, \mathcal{O}_{\mathbb{P}^3}(\mathbb{F} - \Delta)) - \chi(\mathbb{P}^3, \mathcal{O}_{\mathbb{P}^3}(\mathbb{K})) && \text{(by 4.9.2).} \end{aligned}$$

(The last equality comes from $\mathcal{O}_{\mathbb{P}^3}(\mathbb{F} - M) = \mathcal{O}_{\mathbb{P}^3}(\mathbb{K})$.) By the Serre duality theorem, however, we have $\chi(\mathbb{P}^3, \mathcal{O}_{\mathbb{P}^3}(\mathbb{K})) = -\chi(\mathbb{P}^3, \mathcal{O}_{\mathbb{P}^3}) = -1$. Hence, $p_a + 1 = \chi(\mathbb{P}^3, \mathcal{O}_{\mathbb{P}^3}(\mathbb{F})) - \chi(\Delta, \mathcal{O}_{\mathbb{P}^3}(\mathbb{F})_\Delta) + 1$ by 4.9.3. Since $\chi(\mathbb{P}^3, \mathcal{O}_{\mathbb{P}^3}(\mathbb{F})) = \dim H^0(\mathbb{P}^3, \mathcal{O}_{\mathbb{P}^3}((n-4)\mathbb{E})) = \dim \mathcal{L}_{n-4} = \binom{n-1}{3}$, we will complete the proof of 4.12 if we can show the following. □

Lemma 4.12.1 $\chi(\Delta, \mathcal{O}_{\mathbb{P}^3}(\mathbb{F})_\Delta) = (n-4)m - \pi(\Delta) + 1$.

(If we denote the non-singular model of $\Delta = \bigcup_j \Delta_j$ (irreducible decomposition) by $\hat{\Delta} = \bigsqcup_j \hat{\Delta}_j$, then $\pi(\Delta) - 1 = \sum_j (\pi(\hat{\Delta}_j) - 1) + 2t$.)

Proof We put:

$$\begin{cases} L\colon \text{a general plane in } \mathbb{P}^3, \\ \mathbb{E} = [L],\ \mathbb{F} = (n-4)\mathbb{E}, \\ \operatorname{Sing}(\Delta) = \{\mathfrak{t}_1, \ldots, \mathfrak{t}_k, \ldots, \mathfrak{t}_t\} \text{ (all of them are triple points of } \Delta (4.5, \text{II})), \\ \hat{\mu}\colon \hat{\Delta} \to \Delta \subset M \subset \mathbb{P}^3 \text{ desingularization}, \\ \hat{\mu}^{-1}(\mathfrak{t}_k) = \{\hat{\mathfrak{t}}_{k1}, \hat{\mathfrak{t}}_{k2}, \hat{\mathfrak{t}}_{k3}\}. \end{cases}$$

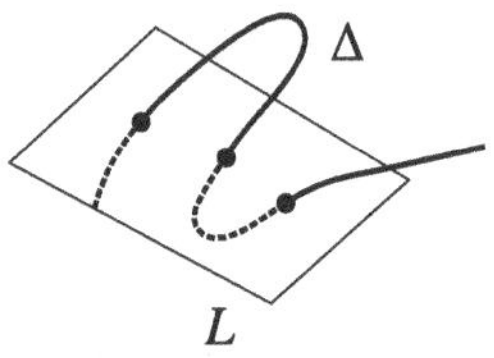

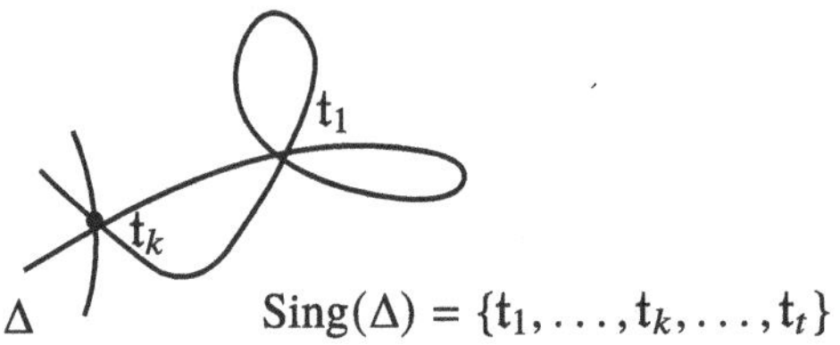

Then the conductor of Δ is $\displaystyle\sum_{k=1}^{t}\sum_{\nu=1}^{3} \hat{\mathfrak{t}}_{k\nu}$. Since we have the natural isomorphism $\displaystyle\bigoplus_{\nu=1}^{3} \mathcal{O}_{\hat{\Delta}}(-\hat{\mathfrak{t}}_{k\nu})_{\hat{\mathfrak{t}}_{k\nu}} \cong \mathcal{O}_\Delta(\mathfrak{t}_k)_{\mathfrak{t}_k}$, we obtain:

(a) $\displaystyle\hat{\mu}_*(\mathcal{O}_{\hat{\Delta}}(\hat{\mu}^*\mathbb{F}_\Delta - \sum_{k=1}^{t}\sum_{\nu=1}^{3} \hat{\mathfrak{t}}_{k\nu})) \overset{\sim}{\leftarrow} \mathcal{O}_{\mathbb{P}^3}(\mathbb{F} - \sum_{k=1}^{t} \mathfrak{t}_k)_\Delta$,
where $\mathbb{F}_\Delta$ denotes the restriction of $\mathbb{F}$ to Δ and $\mathcal{O}_{\mathbb{P}^3}(\mathbb{F} - \sum_k \mathfrak{t}_k)_\Delta = \{\varphi \in \mathcal{O}_{\mathbb{P}^3}(\mathbb{F})_\Delta \mid \varphi(\mathfrak{t}_k) = 0\ (k = 1, 2, \ldots)\}$. Then, we clearly have the exact sequence:

(b) $\displaystyle 0 \to \mathcal{O}_{\mathbb{P}^3}(\mathbb{F} - \sum_{k=1}^{t} \mathfrak{t}_k)_\Delta \to \mathcal{O}_{\mathbb{P}^3}(\mathbb{F})_\Delta \to T \to 0$,
where T is a sheaf over Δ satisfying $\operatorname{Supp}(T) = \operatorname{Sing}(\Delta) = \cup_{k=1}^{t} \mathfrak{t}_k$, $\dim H^0(\Delta, T) = t$ and, hence, $\chi(\Delta, T) = t$. This yields:

(c) $\displaystyle\chi(\Delta, \mathcal{O}_{\mathbb{P}^3}(\mathbb{F})_\Delta) = \chi(\hat{\Delta}, \mathcal{O}_{\hat{\Delta}}(\hat{\mu}^*\mathbb{F}_\Delta - \sum_{k=1}^{t}\sum_{\nu=1}^{3} \hat{\mathfrak{t}}_{k\nu})) + t$.

If $\mathfrak{d} = L \cdot \Delta$ (hence $\deg \mathfrak{d} = m$), then $\mathbb{F}_\Delta = (n-4)[\mathfrak{d}]$. Similarly, if we put $\mathfrak{d}_j = L \cdot \Delta_j$, $\hat{\mathfrak{d}} = \hat{\mu}^*\mathfrak{d}$ and $\hat{\mathfrak{d}}_j = \hat{\mu}^*\mathfrak{d}_j$, then obviously $\hat{\mathfrak{d}} = \sum_j \hat{\mathfrak{d}}_j$. We next distribute the conductor $\mathfrak{a} = \sum_{k=1}^{t}\sum_{\nu=1}^{3} \hat{\mathfrak{t}}_{k\nu}$ to each $\hat{\Delta}_j$ and write $\sum_j \mathfrak{a}_j$ (where $\mathfrak{a}_j$ is the divisor on $\hat{\Delta}_j$, $\hat{\Delta} = \sqcup_j \hat{\Delta}_j$). It is clear that $\deg \mathfrak{a} = 3t$. Noting that $\hat{\Delta} = \sqcup_j \hat{\Delta}_j$ and $\hat{\mu}^*\mathbb{F}_\Delta = (n-4)[\hat{\mathfrak{d}}] = \sum_j (n-4)\hat{\mathfrak{d}}$, we get:

(d) $$\mathscr{O}_{\hat{\Delta}}\Big(\hat{\mu}^*\mathbb{F}_\Delta - \sum_{k=1}^{t}\sum_{\nu=1}^{3} \hat{\mathfrak{t}}_{k\nu}\Big) = \bigoplus_j \mathscr{O}_{\hat{\Delta}_j}((n-4)\hat{\mathfrak{d}} - \mathfrak{a}_j).$$

By the Riemann–Roch theorem for curves, we have

(e) $$\chi(\hat{\Delta}_j, \mathscr{O}_{\hat{\Delta}_j}((n-4)\hat{\mathfrak{d}} - \mathfrak{a}_j)) = (n-4)\deg \hat{\mathfrak{d}}_j - \deg \mathfrak{a}_j - \pi(\hat{\Delta}_j) + 1.$$

From (c), (d), and (e),

$$\begin{aligned}
\chi(\Delta, \mathscr{O}_{\mathbb{P}^3}(\mathbb{F})_\Delta) &= \sum_j \chi(\Delta_j, \mathscr{O}_{\hat{\Delta}_j}((n-4)\hat{\mathfrak{d}} - \mathfrak{a}_j)) + t \\
&= (n-4)\deg \hat{\mathfrak{d}} - \deg \mathfrak{a} - \sum_j (\pi(\hat{\Delta}_j) - 1) + t \\
&= (n-4)m - \pi(\Delta) + 1
\end{aligned}$$

which is what we want. □

4.13 In order to compute the Chern numbers c_1^2, c_2, we prepare things in (I), (II), and (III) below. The actual calculations will be done in (IV).

Let $\lambda\colon S \to M \hookrightarrow \mathbb{P}^3$ be the projection, let (w, x, y, z) denote a system of homogeneous coordinates of $\mathbb{P}^3$ (which is assumed to be taken sufficiently general with respect to M) and let $f(w, x, y, z) = 0$ be the irreducible equation defining M. The degree of f is n.

(I) Hyperplane sections of M (finding the formula for $c_2(S)$).

Put

$$\begin{cases} L_\tau\colon w - \tau z = 0 & (\tau \in \mathbb{P}^1), \\ L_\infty\colon z = 0, \end{cases}$$

$$\begin{cases} C_\tau\colon L_\tau \cap M, \\ \ell\colon w = z = 0 & (\text{of course } \ell \subset L_\tau). \end{cases}$$

For a general τ, we have

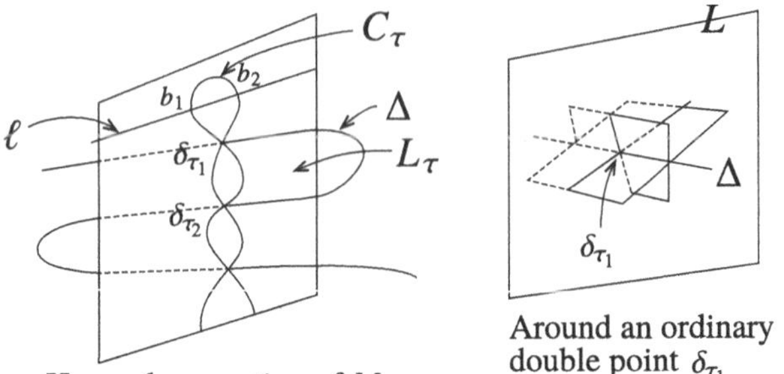

Hyperplane section of M

Around an ordinary double point δ_{τ_1}

$$\begin{cases} L_\tau \cap \Delta = \delta_{\tau,1} + \delta_{\tau,2} + \cdots + \delta_{\tau,m} & (m\text{: degree of } \Delta), \\ \ell \cap M = b_1 + b_2 + \cdots + b_n & (n\text{: degree of } M). \end{cases}$$

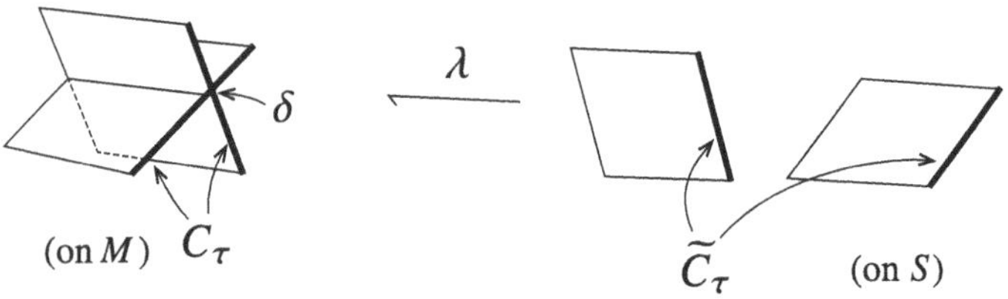

Let $\widetilde{C}_\tau$ be the proper inverse image of C_τ in S, that is, $\widetilde{C}_\tau = \lambda^{-1}[C_\tau]$. Then $\widetilde{C}_\tau$ is non-singular for a general τ. On the other hand, there are a finite number of points $\tau_1, \ldots, \tau_j, \ldots, \tau_c (\in \mathbb{P}^1)$ such that L_{τ_j} contacts M at a point $\mathfrak{o}_j$, that is, c is the number of tangent planes of M passing through ℓ. The number c is known as the *class* of M.

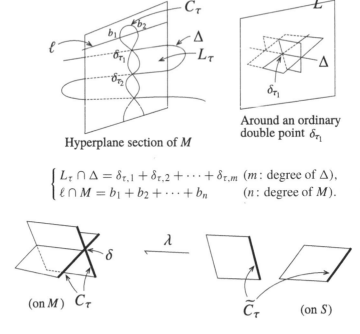

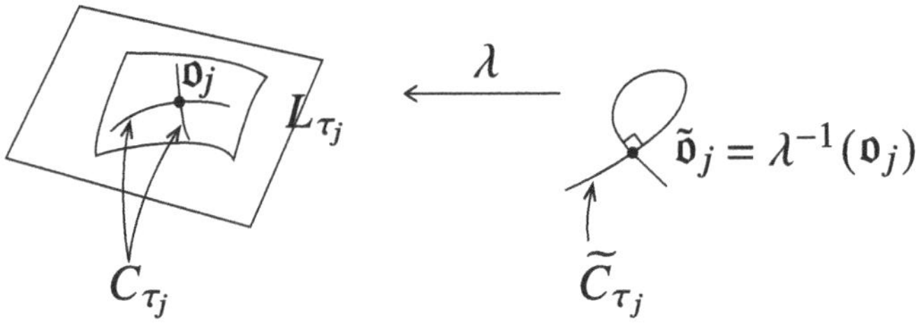

Since (w, x, y, z) is taken generally with respect to M, we can assume $\mathfrak{o}_j \notin \Delta$. Then, $\tilde{\mathfrak{o}}_j = \lambda^{-1}(\mathfrak{o}_j)$ is a node (ordinary double point) of $\widetilde{C}_{\tau_j}$.

4.13.1 If $\tau \neq \tau_j$ $(1 \leqq j \leqq c)$, then $\widetilde{C}_\tau$ is non-singular.

Proof We investigate $\widetilde{C}_\tau$ in a neighborhood of a point in $\lambda^{-1}(\mathfrak{q})$ when L_τ passes through a cuspidal point $\mathfrak{q}$ of M. (In fact, since $\tau \neq \tau_j$, $\mathrm{Sing}(C_\tau) \subset L_\tau \cap \Delta$. Hence in order to study $\mathrm{Sing}(\widetilde{C}_\tau)$, it suffices to work in a neighborhood of $\lambda^{-1}(L_\tau \cap \Delta)$ in $\widetilde{C}_\tau$.) If L_τ passes through a singular point $\mathfrak{p}$ other than a cuspidal point, it is almost clear that $\widetilde{C}_\tau$ is non-singular in a neighborhood of $\lambda^{-1}(\mathfrak{p})$.)

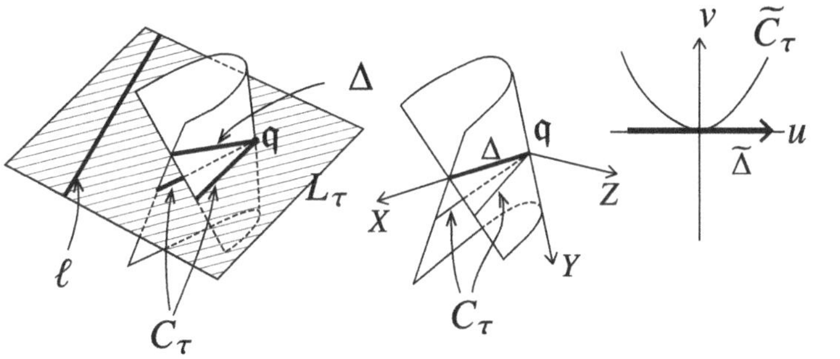

Letting (X, Y, Z) and (u, v) denote local coordinates around $\mathfrak{q}$ and $\tilde{\mathfrak{q}} = \lambda^{-1}(\mathfrak{q})$, respectively, we may write locally

$$\begin{cases} \lambda\colon (u, v) \longrightarrow (X, Y, Z) = (u^2, v, uv), \\ M\colon XY^2 - Z^2 = 0, \\ L_\tau\colon AX + BY + CZ = 0, \\ \widetilde{C}_\tau\colon Au^2 + Bv + Cuv = 0. \end{cases}$$

We may assume that $\mathfrak{q} = (0, 0, 0)$, $A \neq 0$, $B \neq 0$, $C \neq 0$ and, furthermore, $B+Cu \neq 0$ in a neighborhood of $\tilde{\mathfrak{q}} = (0, 0)$. Then, since $\widetilde{C}_\tau\colon (B+Cu)v+Au^2 = 0$, we see that $\tilde{\mathfrak{q}} = (0, 0)$ is a simple point of $\widetilde{C}_\tau$.

Therefore, $\widetilde{C}_\tau$ is non-singular when $\tau \neq \tau_j$ $(1 \leqq j \leqq c)$. □

Definition 4.13.2 For a general τ, we put $g = \pi(\widetilde{C}_\tau)$.

Since C_τ is a plane curve of degree n having m nodes, we have the following:

(a) $2g - 2 = n(n-3) - 2m$.

Now, $S = \bigcup_{\tau \in \mathbb{P}^1} \widetilde{C}_\tau$, $C_\tau \cap C_\lambda = b_1 + b_2 + \cdots + b_n$ $(= \ell \cap M)$. Moreover, we can write $\widetilde{C}_\tau \cap \widetilde{C}_\lambda = \tilde{b}_1 + \tilde{b}_2 + \cdots + \tilde{b}_n$ and assume that $\widetilde{C}_\tau$ and $\widetilde{C}_\lambda$ meet normally at $\tilde{b}_k$.

Since $\{\tilde{b}_k\}_{1 \leqq k \leqq n}$ is the set of all base points of the pencil $\{\widetilde{C}_\tau\}_{\tau \in \mathbb{P}^1}$, in order to have a fibration from it, we perform a quadratic transformation $Q_{\tilde{b}_k}$ with center $\tilde{b}_k$ to S $(1 \leqq k \leqq n)$ to get $\widehat{S}$, that is, $\widehat{S} = Q_{\tilde{b}_1} Q_{\tilde{b}_2} \cdots Q_{\tilde{b}_n}(S)$. Then

$$c_2(\widehat{S}) = \chi(\mathbb{P}^1) \cdot \chi(\widetilde{C}_\tau) + \sum_{j=1}^{c} (\chi(\widetilde{C}_{\tau_j}) - \chi(\widetilde{C}_\tau)),$$

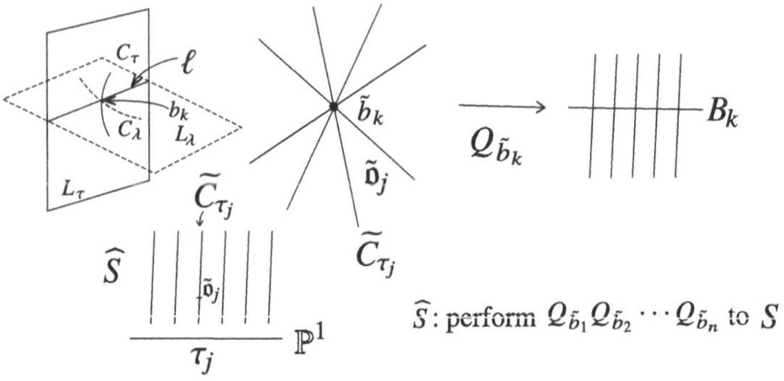

where $\chi(V)$ denotes the Euler number of V. Since $\tilde{\mathfrak{o}}_j$ is the unique singular point of $\widetilde{C}_{\tau_j}$ which is a node, we have $\chi(\widetilde{C}_{\tau_j}) - \chi(\widetilde{C}_\tau) = 1$. Hence $c_2(\widehat{S}) = 2(2-2g) + c$. On the other hand, since the second Betti number increases by one after a quadratic transformation, we have $c_2(S) = c_2(\widehat{S}) - n$. Hence, we obtain

(b) $c_2(S) = 4 - 4g + c - n.$

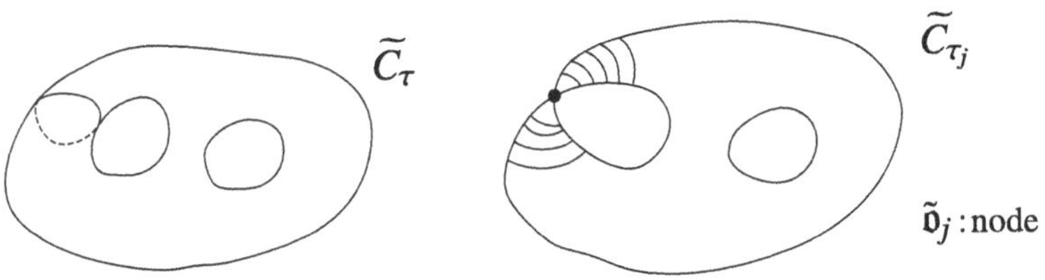

(II) Polar curves.

Recall M: $f(w, x, y, z) = 0$. We put M_x: $\partial_x f = 0$ (where $\partial_x f = \frac{\partial f}{\partial x}$, polynomial of degree $n-1$). Since we have $\partial_x f(P) = 0$ when $P \in \Delta$, there is a curve D_x satisfying $M_x \cap M = \Delta \cup D_x$.

Definition 4.13.3 D_x is called a polar curve of M.

Now let $(\zeta_0, \zeta_1, \zeta_2, \zeta_3)$ be a system of homogeneous coordinates on $\mathbb{P}^3$ and write

$$\zeta_\nu = a_{\nu 0} w + a_{\nu 1} x + a_{\nu 2} y + a_{\nu 3} z.$$

We regard $a_{\nu k}$ as parameters. Since $\partial_x f = \sum_{\nu=0}^{3} \frac{\partial \zeta_\nu}{\partial x} \frac{\partial f}{\partial \zeta_\nu} = \sum_{\nu=0}^{3} a_{\nu 1} \frac{\partial f}{\partial \zeta_\nu}$ is linear in the $a_{\nu 1}$'s, $\{D_x\}$ forms a linear system.

4.13.4 The set of base points of $\{D_x\}$ coincides with the set of all cuspidal points $\mathfrak{q}_j$.

Proof

(i) Obviously, $\mathfrak{p} \notin \Delta$ is not a base point of $\{D_x\}$. So we consider points of Δ.

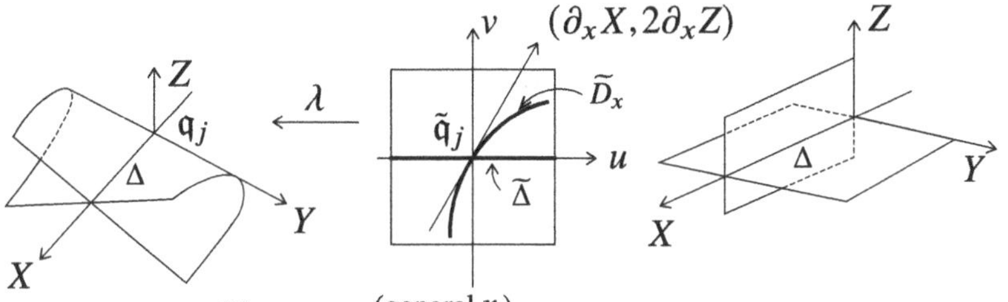

(ii) (general x)

(ii) We consider a cuspidal point $\mathfrak{q}_j$. Recall that λ: $(u, v) \to (X, Y, Z) = (u^2, v, uv)$. We claim that $\mathfrak{q}_j$ is a base point of $\{D_x\}$. This can be seen as follows. Assume that $w \neq 0$ at $\mathfrak{q}_j$. Putting $w = 1$ and considering (x, y, z) as a system of inhomogeneous coordinates on $\mathbb{P}^3$, we have $f(1, x, y, z) = XY^2 - Z^2$. Hence $\partial_x f = (\partial_x X) \cdot Y^2 + (2\partial_x Y) \cdot XY - (2\partial_x Z) \cdot Z$ and we have $\lambda^*(\partial_x f)(u, v) = \partial_x X \cdot v^2 + 2\partial_x Y \cdot u^2 v - 2\partial_x Z \cdot uv = v(\partial_x X \cdot v + \partial_x Y \cdot u^2 + \partial_x Z \cdot u)$. It follows that $\widetilde{\Delta} = \lambda^{-1}[\Delta]$: $v = 0$, $\widetilde{D}_x = \lambda^{-1}[D_x]$: $\partial_x X \cdot v + \partial_x Y \cdot u^2 + \partial_x Z \cdot u = 0$. This implies that every $\widetilde{D}_x$ passes through $\tilde{\mathfrak{q}}_j = (0, 0)$. Hence $\mathfrak{q}_j \in D_x$ for all x. We also remark that $\mathfrak{q}_j$ is a non-singular point of a general D_x. (For a general x, since we have $\partial_x X \neq 0$, $\partial_x Y \neq 0$, $\partial_x Z \neq 0$ at $\mathfrak{q}_j$, the picture of $\widetilde{D}_x$ is as in the above figure.)

(iii) If $\mathfrak{p} \in \Delta$ but $\mathfrak{p} \neq \mathfrak{q}_j$, then $\mathfrak{p} \notin D_x$ for a general x. This can be seen as follows. If $\mathfrak{p}$ is a double point, then taking coordinates as in (ii), we have $f(1, x, y, z) = Y \cdot Z$. Hence $\partial_x f = (\partial_x Y) \cdot Z + (\partial_x Z) \cdot Y$. For a general x, we can assume that $\partial_x Y \neq 0$ and $\partial_x Z \neq 0$ in a neighborhood of $\mathfrak{p}$. Therefore, if we consider $\partial_x f = 0$, then $Y = 0 \Leftrightarrow Z = 0$. This implies that $M_x \cap M = \Delta$ in a neighborhood of $\mathfrak{p}$, and D_x does not pass through $\mathfrak{p}$. If $\mathfrak{p}$ is a triple point, then we can write $f(1, x, y, z) = XYZ$ and argue similarly.

By (i), (ii), and (iii), the set of base points of $\{D_x\}$ coincides with the set of all cuspidal points $\{\mathfrak{q}_j\}$. □

$\{D_x\}$ does not have a base point outside $\{\mathfrak{q}_j\}$. Moreover, as we have already seen, $\mathfrak{q}_j$ is a simple point of a general D_x. It follows from Bertini's theorem that a general D_x is non-singular.

Now, we let L be a plane in $\mathbb{P}^3$ and put $\mathbb{E} = [L]$. Then $[M_x] = (n-1)\mathbb{E}$ by definition. On the other hand, since we have $M_x \bigcap M = \Delta \bigcap D_x$, we obtain

(c) $[\widetilde{\Delta} + \widetilde{D}_x] = (n-1)\widetilde{\mathbb{E}}$ (where $\widetilde{\mathbb{E}} = \lambda^*\mathbb{E}$).

Next, we fix a general x and study the intersection points of D_x and Δ. From the above discussion, we know that $D_x \cap \Delta$ contains all the cuspidal points $\mathfrak{q}_1, \ldots, \mathfrak{q}_\gamma$ of M. At a general point of Δ, $f(1, x, y, z) = YZ$, $f_x = Y_x Z + Z_x Y = 0$ and, hence $(\partial_x f)_{Z=0} = Z_x Y$, $(\partial_x f)_{Y=0} = Y_x Z$. Then for $\mathfrak{p} \in D_x \cap \Delta \subset M_x \cap M$, we have either $Z_x(\mathfrak{p}) = 0$, $Y_x(\mathfrak{p}) \neq 0$ or $Z_x(\mathfrak{p}) \neq 0$, $Y_x(\mathfrak{p}) = 0$. In the former case, for example, a neighborhood of $\mathfrak{p}$ is Δ on $Y = 0$, and $\Delta + D_x$ on $Z = 0$.

Therefore, if we let $\mathfrak{r}_1, \dots, \mathfrak{r}_\sigma$ be the points of intersection of $Z_x Y_x = 0$ and Δ, then we can write $D_x \cap \Delta = \mathfrak{q}_1 + \cdots + \mathfrak{q}_\gamma + \mathfrak{r}_1 + \cdots + \mathfrak{r}_\sigma$ and $\widetilde{D}_x \cap \widetilde{\Delta} = \tilde{\mathfrak{q}}_1 + \cdots + \tilde{\mathfrak{q}}_\gamma + \tilde{\mathfrak{r}}_1 + \cdots + \tilde{\mathfrak{r}}_\delta$ (the latter follows from the former). By Bertini's theorem, $\tilde{\mathfrak{r}}_k$ is a simple point of $\widetilde{D}_x$. We obtain

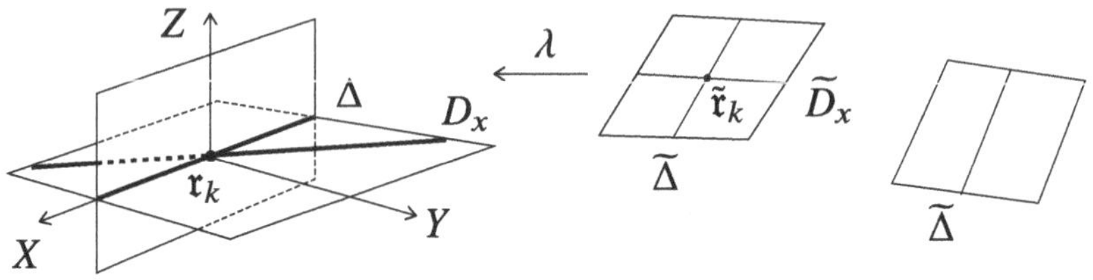

(d) $\widetilde{D}_x \cdot \widetilde{\Delta} = \gamma + \sigma.$

(III) Second polar curves.

We put $M_{xx} : \partial_x^2 f = 0$. M_{xx} is of degree $n - 2$.

(i) At a cuspidal point $\mathfrak{q}_j \in \Delta$, we have $f = f(1, x, y, z) = XY^2 - Z^2$ and, since (w, x, y, z) can be chosen generally, we can assume that $f_{xx}(\mathfrak{q}_j) = -2Z_x^2 \neq 0$, i.e., $M_{xx} \not\ni \mathfrak{q}_j$.

(ii) At a triple point $\mathfrak{t}_i \in \Delta$, we have $f = XYZ$ and $\partial_x^2 f = 2\partial_x X \partial_x Y \cdot Z + \partial_x Y \partial_x Z \cdot X + \partial_x Z \partial_x X \cdot Y + \partial_x^2 X \cdot YZ + \cdots$, and we can assume that $\partial_x X \partial_x Y \partial_x Z \neq 0$ at $\mathfrak{t}_i$. Then M_{xx} and Δ meet normally at $\mathfrak{t}_i$.

(iii) At a general point $\mathfrak{p} \in \Delta$, we have $f = YZ$ and $f_{xx} = Y_x Z_x + Y_{xx} Z + Z_{xx} Y$. Hence the necessary and sufficient condition to have $\mathfrak{p} \in M_{xx} \cap \Delta$ is $f_{xx}(\mathfrak{p}) = Y_x(\mathfrak{p}) Z_x(\mathfrak{p}) = 0$, that is, $\mathfrak{p} = \mathfrak{r}_i$. On the other hand, since D_x and Δ meet normally at $\mathfrak{r}_i$, we have $Z_{xx}(\mathfrak{r}_i) \neq 0$ and, hence, M_{xx} and Δ meet normally at $\mathfrak{r}_i$.

In sum,

$$M_{xx} \cap \Delta = \mathfrak{r}_1 + \cdots + \mathfrak{r}_\sigma + \mathfrak{t}_1 + \cdots + \mathfrak{t}_t$$

and, at each $\mathfrak{r}_i$ or $\mathfrak{t}_i$, M_{xx} has a simple point and meets every component of Δ normally.

(e) $M_{xx} \cdot \Delta = \sigma + 3t.$

(IV) Finally, we consider another polar curve $D_y : \partial_y f = 0$.

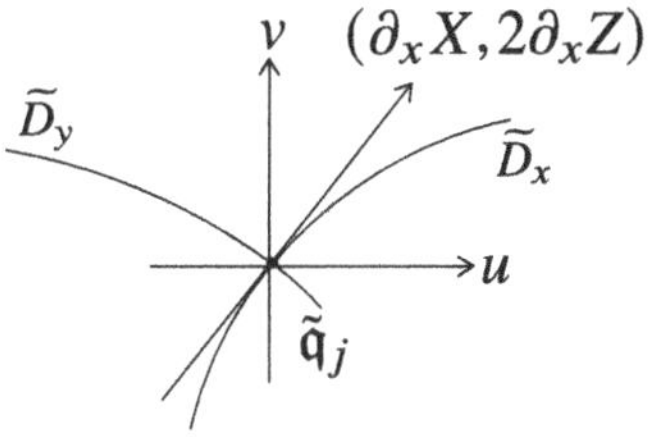

We have $D_x \cap D_y \cap \Delta = \bigcup \mathfrak{q}_i$. We consider a point $\mathfrak{p} \in D_x \cap D_y$ with $\mathfrak{p} \notin \Delta$. That is, we assume that $f_x(\mathfrak{p}) = f_y(\mathfrak{p}) = 0$ and $\mathfrak{p} \notin \Delta$. Since the tangent plane $T_p(M)$ at a smooth point p is given by

$$T_p(M) \colon f_w(p)w + f_x(p)x + f_y(p)y + f_z(p)z = 0,$$

we have $T_{\mathfrak{p}}(M) = L_\tau$ at $\mathfrak{p} \in D_x \cap D_y$, where $\tau = -\frac{f_z(\mathfrak{p})}{f_w(\mathfrak{p})}$. (Recall that $L_\tau : w - \tau z = 0$ and that $\mathfrak{o}_j$ is the point at which L_{τ_j} has a contact to M in the notation of (I).) Hence $\mathfrak{p} = \mathfrak{o}_j$ for some j. Conversely, when $\mathfrak{p} = \mathfrak{o}_j$, then $\mathfrak{o}_j \in D_x \cap D_y$, since $f_x(\mathfrak{o}_j) = f_y(\mathfrak{o}_j) = 0$. Hence we obtain

$$\widetilde{D}_x \cap \widetilde{D}_y = \tilde{\mathfrak{q}}_1 + \cdots + \tilde{\mathfrak{q}}_\gamma + \tilde{\mathfrak{o}}_1 + \cdots + \tilde{\mathfrak{o}}_c$$

and

(f) $\widetilde{D}_x^2 = \widetilde{D}_x \cdot \widetilde{D}_y = \gamma + c.$

Recall that we have $\mathbb{E} = [L_\tau]$, $\widetilde{\mathbb{E}} = \lambda^*\mathbb{E} = [\widetilde{C}_\tau]$, $\widetilde{\mathbb{E}}^2 = \widetilde{C}_\tau^2 = \widetilde{C}_\tau \cdot \widetilde{C}_{\tau'} = n$, $\widetilde{E}^2 + K \cdot \widetilde{\mathbb{E}} = \widetilde{C}_\tau^2 + K \cdot \widetilde{C}_\tau = 2\pi(\widetilde{C}_\tau) - 2 = 2g - 2$. Then it follows from **(a)** that

(g) $\widetilde{\mathbb{E}}^2 = n,$
(h) $K \cdot \widetilde{\mathbb{E}} = n(n-4) - 2m.$

Now, we infer from 4.9.1 and **(c)** that

$$[\widetilde{\Delta}] = (n-4)\widetilde{\mathbb{E}} - K,$$
$$[\widetilde{D}_x] = (n-1)\widetilde{\mathbb{E}} - [\widetilde{\Delta}] = 3\widetilde{\mathbb{E}} + K.$$

These together with **(d)** and **(h)** give us

$$\begin{aligned}\gamma + \sigma = \widetilde{D}_x \cdot \widetilde{\Delta} &= (3\widetilde{\mathbb{E}} + K)((n-4)\widetilde{\mathbb{E}} - K)\\ &= 3n(n-4) + (n-7)K\widetilde{\mathbb{E}} - K^2\\ &= n(n^2 - 8n + 16) - (2n - 14)m - c_1^2.\end{aligned}$$

We also have $\sigma = (n-2)m - 3t$ from **(e)**, since the degrees of M_{xx} and Δ are $n-2$ and m, respectively. Hence

(i) $c_1^2 = n(n-4)^2 - (3n-16)m + 3t - \gamma$.

Next, it follows from **(f)**, **(g)**, and **(h)** that

$$\gamma + c = \widetilde{D}_x^2 = (3\widetilde{\mathbb{E}} + K) = n(6n-15) - 12m + c_1^2.$$

Hence, from this and **(a)**, **(b)**, and **(i)**,

(j) $c_2 = n(n^2 - 4n + 6) - (3n-8)m + 3t - 2\gamma$.

We have only to compute γ.

Let $\widehat{\widetilde{\Delta}}$ be the non-singular model of $\widetilde{\Delta}$. Each $\tilde{\mathfrak{t}}_{i_\nu}$ is a double point of $\widetilde{\Delta}$ and there are $3t$ such points in total. We have $\mathrm{Sing}(\widetilde{\Delta}) = \bigcup_{i_\nu} \tilde{\mathfrak{t}}_{i_\nu}$. Hence $\pi(\widehat{\widetilde{\Delta}}) = \pi(\widetilde{\Delta}) - 3t$. On the other hand, $\widehat{\widetilde{\Delta}} \to \widehat{\Delta}$ is a double covering with branch points $\tilde{\mathfrak{q}}_j (= \widehat{\mathfrak{q}}_j)$'s. Since $\widehat{\Delta} = \sqcup \widehat{\Delta}_j$, we can compute the Euler number of $\widehat{\widetilde{\Delta}}$ as

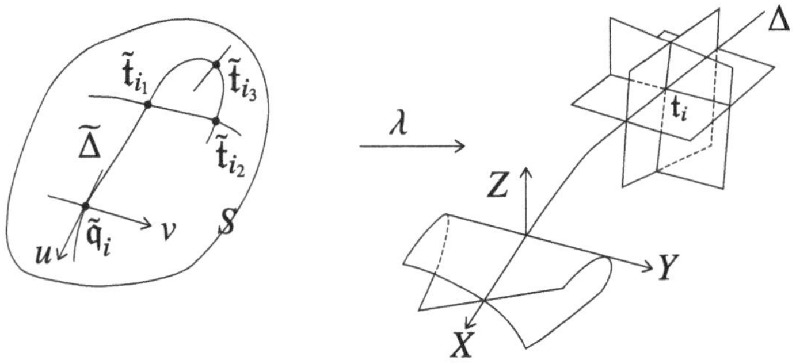

$$2 - 2\pi(\widehat{\widetilde{\Delta}}) = 2(2 - 2\pi(\widehat{\Delta})) - \gamma$$
$$= 2\sum_j (2 - 2\pi(\widehat{\Delta}_j)) - \gamma.$$

From this and $\pi(\widehat{\widetilde{\Delta}}) = \pi(\widetilde{\Delta}) - 3t$, we get

$$-\gamma = 2\sum_j (2\pi(\widehat{\Delta}_j) - 2) - 2\pi(\widetilde{\Delta}) + 2 + 6t.$$

Since $2\pi(\widetilde{\Delta}) - 2 = (\widetilde{\Delta} + K)\cdot\widetilde{\Delta} = (n-4)\widetilde{\mathbb{E}}\cdot\widetilde{\Delta} = (n-4)\cdot 2m$, $\pi(\Delta) - 1 = \sum_j (\pi(\widehat{\Delta}_j) - 1) + 2t$, we get

(k) $-\gamma = 4\pi(\Delta) - 4 - (2n-8)m - 2t$.

Consequently, we infer from **(i)** and **(j)** that

(l) $c_1^2 = n(n-4)^2 - (5n-24)m + t + 4\pi(\Delta) - 4,$

(m) $c_2 = n(n^2 - 4n + 6) - (7n - 24)m - t + 8\pi(\Delta) - 8.$

Example 4.14 Enriques surface.

Let (z_0, z_1, z_2, z_3) be a system of homogeneous coordinates on $\mathbb{P}^3$ and put

$$M\colon f(z_0, z_1, z_2, z_3) = a_0(z_0z_1z_2)^2 + a_1(z_1z_2z_3)^2 + a_2(z_2z_3z_0)^2 + a_3(z_3z_0z_1)^2 + z_0z_1z_2z_3\psi(z_0, z_1, z_2, z_3) = 0,$$

where ψ is a quadratic form.

Definition 4.14.1 The non-singular model S of a general M is called an Enriques surface.

4.14.2 An Enriques surface is the first example of a non-rational surface with $p_a = p_g = 0$.

First of all, we investigate $\Delta = \mathrm{Sing}(M)$ for a general M. Varying a_i and ψ, we get a linear system $\Lambda = \{M\}$. Its general member M has no singular points outside the base locus by Bertini's theorem.

Putting $\Delta_{\lambda\nu}\colon z_\lambda = z_\nu = 0$, the set of all base points of $\{M\}$ clearly coincides with $\bigcup_{0\leqq\lambda<\nu\leqq 3} \Delta_{\lambda\nu}$. In the following (i) and (ii), we will show that $\Delta = \bigcup_{0\leqq\lambda<\nu\leqq 3} \Delta_{\lambda\nu}$ for M general.

After a suitable change of coordinates $(z_0, \cdots, z_3) \to (c_0z_0, \ldots, c_3z_3)$, we may assume that $M\colon f = (z_0z_1z_2)^2 + (z_1z_2z_3)^2 + (z_2z_3z_0)^2 + (z_3z_0z_1)^2 + z_0z_1z_2z_3\psi = 0$, where ψ is a general quadratic form.

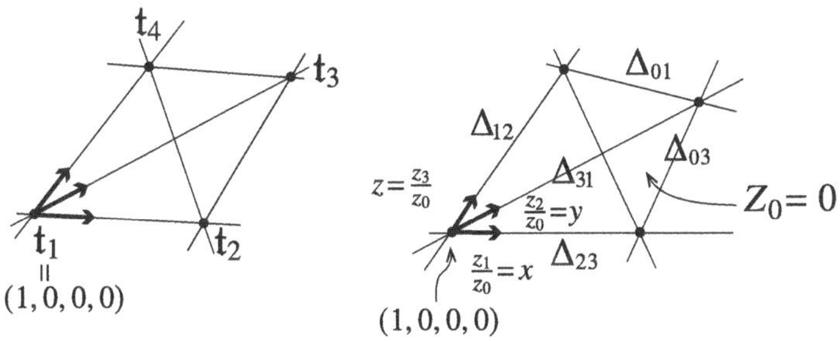

We put $x = \frac{z_1}{z_0}$, $y = \frac{z_2}{z_0}$, $z = \frac{z_3}{z_0}$.

(i) $\mathrm{t}_i = (0, \ldots, 1, \ldots, 0)$ $(1 \leqq i \leqq 4)$ are triple points.

We study the singular points near $\mathrm{t}_1 = (1, 0, 0, 0)$. In a neighborhood of t_1, $M\colon f = x^2y^2z^2 + y^2z^2 + z^2x^2 + x^2y^2 + xyz\psi(1, x, y, z) = 0$ $(\psi(1, 0, 0, 0) \neq 0)$. Putting $\alpha = \psi + xyz$ $(\alpha(\mathrm{t}_1) \neq 0)$, we work with a new system of coordinates $(\xi, \eta, \zeta) = (\frac{x}{\alpha}, \frac{y}{\alpha}, \frac{z}{\alpha})$. Since $f = \alpha xyz + y^2z^2 + z^2x^2 + x^2y^2 = 0$, multiplying it by α^{-4}, we get

$$M: \xi\eta\zeta + \eta^2\zeta^2 + \zeta^2\xi^2 + \xi^2\eta^2 = 0$$

in a neighborhood of $\mathfrak{t}_1$. We define (X, Y, Z) by $(\xi, \eta, \zeta) = (\frac{X}{A}, \frac{Y}{A}, \frac{Z}{A})$, where $A = 1 + X^2 + Y^2 + Z^2 + XYZ$. Then, since

$$(X + YZ)(Y + ZX)(Z + XY) = AXYZ + Y^2Z^2 + Z^2X^2 + X^2Y^2,$$

considering the result of the change of coordinates $(\xi, \eta, \zeta) = (\frac{x}{\alpha}, \frac{y}{\alpha}, \frac{z}{\alpha})$), we get $f(1, x, y, z) = (X + YZ)(Y + ZX)(Z + XY)$. Then it is obvious that $\mathfrak{t}_1$ is a triple point. We can argue similarly for the other $\mathfrak{t}_i$'s, and conclude that the number of triple points is $t = 4$.

(ii) Each point of $\Delta \setminus \{\mathfrak{t}_1, \mathfrak{t}_2, \mathfrak{t}_3, \mathfrak{t}_4\}$ is a double point of M.

For example, we study the singular points along the x-axis.
In a neighborhood of $\mathfrak{p} = (1, p, 0, 0)$ $(p \neq 0)$, the equation f of M is

$$\begin{aligned} f &= x^2y^2 + (yz + x\psi + x^2yz)yz + x^2z^2 \\ &= A\left(y + \frac{B + \sqrt{B^2 - AC}}{A}z\right)\left(y + \frac{B - \sqrt{B^2 - AC}}{C}z\right), \end{aligned}$$

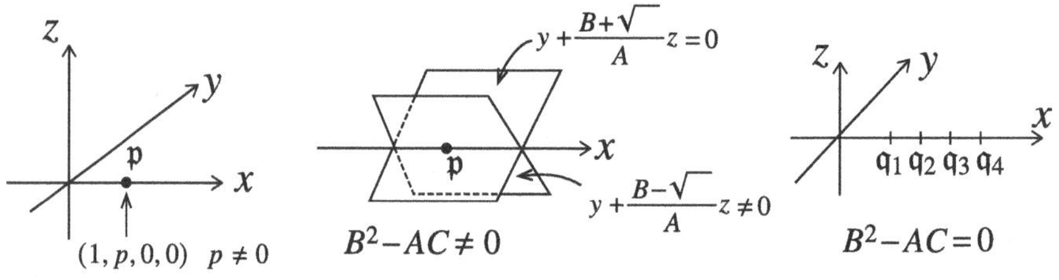

where $A = x^2$, $2B = yz + x\psi + x^2yz$, $C = x^2$ ($A \neq 0$ in a neighborhood of $\mathfrak{p}$). We consider two cases (a), (b) according to the value of $B^2 - AC$.

(a) If $(B^2 - AC)(\mathfrak{p}) \neq 0$, then $\mathfrak{p}$ is an ordinary double point, because $\sqrt{B^2 - AC}$ is holomorphic and $\frac{B+\sqrt{B^2-AC}}{A} \neq \frac{B-\sqrt{B^2-AC}}{A}$.

(b) There are four points $\mathfrak{p}$ on the x-axis (but $x \neq 0$) satisfying $(B^2 - AC)(\mathfrak{p}) = 0$. We put them $\mathfrak{q}_1, \mathfrak{q}_2, \mathfrak{q}_3, \mathfrak{q}_4$. This is because $(B^2 - AC)(1, x, 0, 0) = x^2(\psi(1, x, 0, 0)^2 - x^2)$ and $\psi(1, x, 0, 0)^2 - x^2 = 0$ is a quartic equation in x which has four distinct roots giving $\mathfrak{q}_1$, $\mathfrak{q}_2$, $\mathfrak{q}_3$ and $\mathfrak{q}_4$. We claim that they are cuspidal points. If we use the new system of coordinates $(X, Y, Z) = \left(\frac{B^2-AC}{A^2}, z, y + \frac{B}{A}z\right)$, then $\mathfrak{q}_i$ can be considered as the origin and

$$f = A(Z + \sqrt{X}Y)(Z - \sqrt{X}Y) = A(Z^2 - XY^2) = 0.$$

This implies that $\mathfrak{q}_i$ is a cuspidal point. Considering along every axis similarly, we see that the number of cuspidal points is $\gamma = 4 \times 6 = 24$ in total.

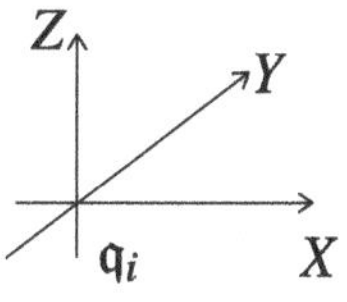

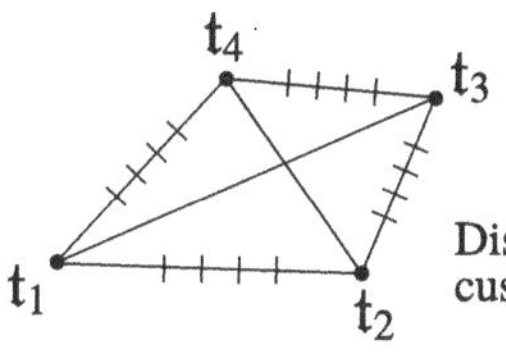

Distribution of cuspidal points

(iii) From the above, we see the following.

M has only ordinary singularity (therefore, we can use formulas in 4.6 to compute p_a, p_g, c_1^2, c_2, etc. of the non-singular model S of M).

$$
\begin{aligned}
&t = 4 && \text{(number of triple points)},\\
&\gamma = 24 && \text{(number of cuspidal points)},\\
&n = 6 && \text{(the degree of } M),\\
&m = 6 && \text{(the degree of } \Delta),\\
&\pi(\Delta_{\lambda\nu}) = 0 \text{ hence } \pi(\Delta) - 1 = \textstyle\sum_{\lambda<\nu}(\pi(\Delta_{\lambda\nu}) - 1) + 2t = 2.
\end{aligned}
$$

We insert these values to 4.6 and 4.11 to get

$$
\begin{aligned}
&p_a = 0, && c_1^2 = 0, && c_2 = 12, && p_g = 0,\\
&q = p_g - p_a = 0, && b_1 = 2q = 0, && b_2 = 10.
\end{aligned}
$$

For the calculations of these, we do not need the value of γ. But we can compute it as $\gamma = 2(n-4)m + 2t - 4\pi(\Delta) + 4 = 24$.

Example 4.15 Let the equation of M in $\mathbb{P}^3$ be

$$M\colon\ f = Ag^2 + 2Bgh + Ch^2 = 0,$$

where g, h, A, B, C are general homogeneous polynomials of respective degree r, $s, n-2r, n-r-s, n-2s$ ($r \geqq s, n \geq 2r, n \geqq r+s+1$), and the degree of f is n.

We investigate $\Delta = \mathrm{Sing}(M)$. As in 4.14, we employ Bertini's theorem to know that

$$\Delta\colon\ g = h = 0,$$

where Δ is a non-singular irreducible curve and $\pi(\Delta) - 1 = \frac{1}{2}rs(r + s - 4)$. We consider $\mathfrak{p} \in \Delta$. Since g, h, A, C are general, A and C do not vanish at $\mathfrak{p}$ simultaneously. We assume that $A(\mathfrak{p}) \neq 0$. Then in a neighborhood of $\mathfrak{p}$,

$$M\colon f = A\left(g + \frac{\sqrt{B^2 - AC}}{A}j\right)\left(g + \frac{B - \sqrt{B^2 - AC}}{A}h\right) = 0.$$

Similarly to 4.14,

(a) If $(B^2 - AC)(\mathfrak{p}) \neq 0$, then $\mathfrak{p}$ is an ordinary double point.
(b) If $(B^2 - AC)(\mathfrak{p}) = 0$, then $\mathfrak{p}$ is a cuspidal point.

On the other hand, considering

$$\begin{cases} \text{degree of } B^2 - AC\colon & 2(n - r - s), \\ \text{degree of } \Delta\colon & m = rs, \end{cases}$$

we have $\gamma = 2rs(n - r - s)$ and $t = 0$. M has only ordinary singularity and we can calculate p_a, c_1^2, c_2, p_g using 4.6 and 4.11:

$$\begin{aligned} p_a &= \binom{n-1}{3} - (n-4)rs + \frac{1}{2}rs(r+s-4), \\ c_1^2 &= n(n-4)^2 - (5n-24)rs + 2rs(r+s-4), \\ c_2 &= n(n^2 - 4n + 6) - (7n - 24)rs + 4rs(r+s-4), \\ p_g &= \binom{n-1-r}{3} + \binom{n-1-s}{3} - \binom{n-r-s-1}{3}. \end{aligned}$$

From these, one can check that $p_a = p_g$ and thus

$$q = p_g - p_a = 0.$$

We calculate p_g using 4.11:

$$p_g = \dim \mathscr{L}_{n-4}(-\Delta) = \dim\{\varphi \mid \varphi \text{ is of degree } n-4 \text{ and } \varphi = 0 \text{ on } \Delta\}$$

For such a φ, we can find $\xi \in \mathscr{L}_{n-r-4}$ and $\eta \in \mathscr{L}_{n-s-4}$ such that $\varphi = \xi g + \eta h$ by Noether's theorem and, therefore, we have the following exact sequence:

$$0 \to \mathscr{L}_{n-4-r-s} \overset{\alpha}{\to} \mathscr{L}_{n-r-4} \bigoplus \mathscr{L}_{n-4-s} \overset{\beta}{\to} \mathscr{L}_{n-4}(-\Delta) \to 0,$$

where $\alpha\colon \tau \to (\xi, \eta) = (h\tau, -g\tau)$, $\beta\colon (\xi, \eta) \to \varphi = \xi g + \eta h$. From this, we get $\dim \mathscr{L}_{n-4}(-\Delta) = \dim \mathscr{L}_{n-r-4} + \dim \mathscr{L}_{n-s-4} - \dim \mathscr{L}_{n-4-r-s}$. Then we get the formula for p_g by $\dim \mathscr{L}_\nu = \binom{\nu+3}{3}$.

Problem 4.16 If we are explicitly given the equation f of M, then we can calculate $q = \dim H^1(S, \mathscr{O})$ and $p_g = \dim H^2(S, \mathscr{O})$ by using formulae 4.6 and 4.11 for the non-singular model S. For the sheaf of germs of holomorphic tangent vector fields Θ, find the formula computing $\dim H^\nu(S, \Theta)$.

Chapter 2
Pluri-Canonical Systems on Algebraic Surfaces of General Type

Abstract After preparing the vanishing theorems, the main Theorem 8 will be proved, that is, it is shown that the m-canonical system is free from base points when $m \geqq 4$, and the m-canonical map is birational when $m \geqq 6$.

2.5 Notation

5.0 Throughout the chapter, we employ the following notation:

- S: a non-singular algebraic surface.
- K: the canonical bundle over S or a canonical divisor.
- x, y, z: points on S.
- $C, C_1, C_2, \ldots, \Theta$: irreducible curves on S.
- X, Y, D: divisors on S.
- $m, n, h, i, j, k \in \mathbb{Z}$.
- When $D = \sum n_i C_i$, we mean that C_i is an irreducible curve and $n_i \in \mathbb{Z}$,

$$D > 0 \Longleftrightarrow n_i > 0.$$

- F: a complex line bundle over S.
  $\mathcal{O}(F)$: the sheaf over S of germs of holomorphic sections of F.
- $\mathcal{O}(F - C) = \{\varphi \in \mathcal{O}(F) \mid \varphi = 0 \text{ on } C\} \cong \mathcal{O}(F - [C])$.
- $\mathcal{O}(F - x - y - \cdots - z) = \{\varphi \in \mathcal{O}(F) \mid \varphi(x) = \varphi(y) = \cdots = \varphi(z) = 0\}$.

We extend the last notation to those involving multiplicities of points.

Take a suitable open covering $S = \bigcup_j U_j$ and put $F = \{f_{jk}(z)\}$. A holomorphic section φ of F over an open subset V of S can be represented by a collection $\{\varphi_j(z)\}$ of holomorphic functions satisfying $\varphi\colon z \to (z, \varphi_j(z))$ on $U_j \cap V$, and $\varphi_j(z) = f_{jk}(z)\varphi_k(z)$ on $U_j \cap U_k \cap V$.

K. Kodaira, *Theory of Algebraic Surfaces*, SpringerBriefs in Mathematics,
https://doi.org/10.1007/978-981-15-7380-4_2

Now, let (z_1, z_2) be a system of local coordinates with the center $x \in U_j \subset S$. If

$$\varphi_j(z) = \sum_{m+n \geqq h} a_{mn} z_1^m z_2^n \qquad (a_{mn} \neq 0 \text{ for some } (m,n), m+n=h),$$

then we say that x *is a zero of order* h of φ.

- $\mathcal{O}(F - hx - ky - \cdots) = \{\varphi \in \mathcal{O}(F) \mid \varphi$ has a zero of order $\geqq h, \geqq k, \ldots$ at $x, y, \ldots\}$.

  Therefore,

$$\begin{cases} \mathcal{O}(F - hx - ky)_z = \mathcal{O}(F)_z & (z \neq x, z \neq y) \\ \mathcal{O}(F)_x / \mathcal{O}(F - hx - ky)_x \cong \mathbb{C}^{\frac{1}{2}h(h+1)}. \end{cases}$$

Definition 5.0.1 $\mathbb{C}_x^m$ means the sheaf satisfying $(\mathbb{C}_x^m)_z = \begin{cases} \mathbb{C}^m & (z = x) \\ 0 & (z \neq x). \end{cases}$ Hence,

5.0.2 $\mathcal{O}(F)/\mathcal{O}(F - hx - ky) \cong \mathbb{C}_x^{\frac{1}{2}h(h+1)} \bigoplus \mathbb{C}_y^{\frac{1}{2}k(k+1)}$.

5.0.3 Let $D = \sum n_i C_i > 0$. On each U_j, we can write $D = (\psi_j)$ with a holomorphic function ψ_j. Then we infer from 2.4 that

$$\mathcal{O}(F - D) = \{\varphi \in \mathcal{O}(F) \mid \varphi_j/\psi_j \text{ is holomorphic}\}.$$

Such a φ is said to be *divisible by D*.

Definition 5.0.4

(a) $\mathcal{O}(F - D - hx - ky - \cdots) = \mathcal{O}(F - D) \cap \mathcal{O}(F - hx - ky - \cdots) = \{\varphi \in \mathcal{O}(F) \mid \varphi$ is divisible by D and has a zero of order $\geqq h, \geqq k, \ldots$ at $x, y, \ldots\}$.

(b) $\mathcal{O}(F - hx - ky - \cdots)_C = \mathcal{O}(F - hx - ky - \cdots)/\mathcal{O}(F - C - hx - ky - \cdots)$ is the restriction of $\mathcal{O}(F - hx - ky - \cdots)$ to C.

Definition 5.0.5 For $D = \sum n_i C_i > 0$, if we take a suitable $\psi \in H^0(S, \mathcal{O}([D]))$, then we can write $D = (\psi)$ and

$$x \in D \Longleftrightarrow \psi(x) = 0 \quad \left(\Longleftrightarrow x \in \bigcup_i C_i\right).$$

Then the order of zero of ψ at x is called the *multiplicity of D at x*. According to whether x is a simple or multiple zero of ψ, x is called a simple or multiple point of D.

The following may be clear.

Proposition 5.0.6 *If x is a point of D of multiplicity $m \geqq h$, then*

$$\mathcal{O}(F - D - hx - ky - \cdots) = \mathcal{O}(F - D - ky - \cdots).$$

Definition 5.1 Recall that $|F| = \{D = (\varphi) \mid \varphi \in H^0(S, \mathcal{O}(F)), \varphi \neq 0\}$. If x is contained in all $D \in |F|$, then we call x a *base point* of $|F|$.

5.1.1 x is a base point of $|F| \iff \varphi(x) = 0$ for all $\varphi \in H^0(S, \mathcal{O}(F))$.

2.6 Vanishing Theorems

Notation 6.0 $h^+ = \max\{h, 0\}$.

Theorem 6.1 *Let m, n be the respective multiplicities of C at $x, y \in C$. If*

$$FC - C^2 - KC > (h - m + 1)^+ m + (k - n + 1)^+ n,$$

then $H^1(C, \mathcal{O}(F - hx - ky)_C) = 0$.

Proof

(a) The case where C is non-singular: We have $m = n = 1$ and $\mathcal{O}(F - hx - ky)_C = \mathcal{O}_C(F_C - hx - ky)$. Letting $\mathfrak{k}$ be the canonical bundle of C, it follows from the Serre duality theorem that $H^1(C, \mathcal{O}_C(F_C - hx - ky)) \cong H^0(C, \mathcal{O}_C(\mathfrak{k} - F_C + [hx] + [ky]))$. We put $\mathfrak{f} = \mathfrak{k} - F_C + [hx] + [ky]$. Applying 3.9 to $\mathfrak{k}$, we get $\mathfrak{f} = K_C + [C]_C - F_C + [hx] + [ky]$. Hence $c(\mathfrak{f}) = KC + C^2 - FC + h + k$. We have $c(\mathfrak{f}) < 0$ by the assumption, since $h + k = (k - m + 1)^+ m + (k - n + 1)^+ n$. Hence $H^0(C, \mathcal{O}_C(\mathfrak{f})) = 0$ and we get $H^1(C, \mathcal{O}(F - hx - ky)_C) = 0$.

(b) In general, let $\mu\colon \widetilde{C} \to C$ be the desingularization and put $\mu^{-1}(x) = \{\mathfrak{p}_1, \ldots, \mathfrak{p}_\lambda, \ldots, \mathfrak{p}_r\}$. We take a local parameter t_λ of $\widetilde{C}$ with the center $\mathfrak{p}_\lambda$ satisfying

$$\mu\colon t_\lambda \to (P_\lambda(t_\lambda), t_\lambda^{m_\lambda}) \qquad (P_\lambda(t_\lambda) \in t_\lambda^{m_\lambda}\mathbb{C}\{t_\lambda\})$$

as in 1.2.

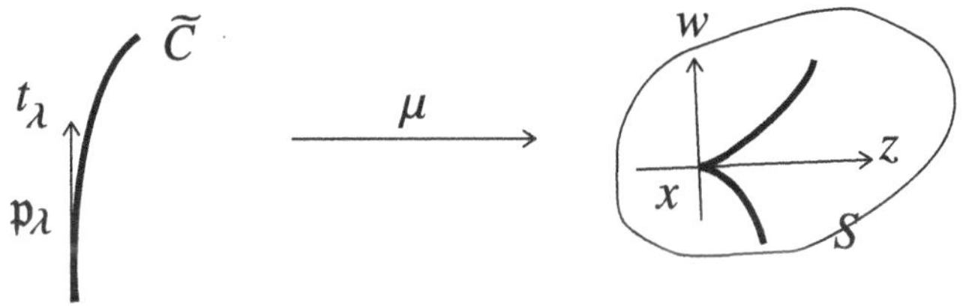

Let the symbols such as $R(w,z) = \prod_{\lambda=1}^{r} R_\lambda(w,z)$, $\sigma_\lambda = t_\lambda^{-c_\lambda}(a_{\lambda 0} + a_{\lambda 1}t_\lambda + \cdots)$ $(a_{\lambda 0} \neq 0)$, and the conductor $\mathfrak{c} = \sum_{\text{singular points}} \sum_\lambda c_\lambda \mathfrak{p}_\lambda$, etc. have the same meaning as in 1.2 and 1.3.1. We put $\mathfrak{d}_x = \sum_{\lambda=1}^{r}(h - m + 1)^+ m_\lambda \mathfrak{p}_\lambda$. Since $m = \sum_{\lambda=1}^{r} m_\lambda$, we have $\deg \mathfrak{d}_x = (h - m + 1)^+ m$. Defining $\mathfrak{d}_y$ similarly for y, we have $\deg \mathfrak{d}_y = (k - n + 1)^+ n$. Putting $\mathfrak{f}' = \mu^* F - \mathfrak{c} - \mathfrak{d}_x - \mathfrak{d}_y$, we obtain from 6.1.1 below the exact sequence

$$\cdots \to H^1(\widetilde{C}, \mathscr{O}(\mathfrak{f}')) \to H^1(C, \mathscr{O}(F - hx - ky)_C) \to 0.$$

On the other hand, letting $\mathfrak{k}$ denote the canonical bundle over $\widetilde{C}$, we have $H^1(\widetilde{C}, \mathscr{O}(\mathfrak{f}')) \cong H^0(\widetilde{C}, \mathscr{O}(\mathfrak{k} - \mathfrak{f}'))$ by the Serre duality theorem, and $\mathfrak{k} = \mu^*[C] + \mu^* K - [\mathfrak{c}]$ by 3.9. We have $c(\mathfrak{k} - \mathfrak{f}') = C^2 + KC - F \cdot C + \deg \mathfrak{d}_x + \deg \mathfrak{d}_y = -(FC - C^2 - KC) + (h - m + 1)^+ m + (k - n + 1)^+ n < 0$ by hypothesis. Hence $H^0(\widetilde{C}, \mathscr{O}(\mathfrak{k} - \mathfrak{f}')) = H^1(\widetilde{C}, \mathscr{O}(\mathfrak{f}')) = 0$. From the above exact sequence, we get $H^1(C, \mathscr{O}(F - hx - ky)_C) = 0$. □

Lemma 6.1.1 *There exists the following exact sequence on* C:

$$0 \to \mu_*(\mathscr{O}(\mu^* F - \mathfrak{c} - \mathfrak{d}_x - \mathfrak{d}_y)) \to \mathscr{O}(F - hx - ky)_C \to \mathscr{M}'' \to 0, \tag{*}$$

where

$$\mathscr{M}''_z = \begin{cases} \text{a finite-dimensional vector space} & (z \in C \text{ is a singular point}), \\ 0 & (z \in C \text{ is a simple point}). \end{cases}$$

Proof We shall show the existence of the natural homomorphism $\tau : \mu_*(\mathscr{O}(\mu^* F - \mathfrak{c} - \mathfrak{d}_x - \mathfrak{d}_y)) \to \mathscr{O}(F - hx - ky)_C$ and define $\mathscr{M}'' = \mathrm{Coker}(\tau)$. Since

$$\begin{cases} \mathscr{O}(\mu^* F - \mathfrak{c} - \mathfrak{d}_x - \mathfrak{d}_y)_\mathfrak{p} = \mathscr{O}(\mu^* F - \mathfrak{c})_\mathfrak{p} & \text{if } \mu(\mathfrak{p}) \neq x, y, \\ \mathscr{O}(F - hx - ky)_z = \mathscr{O}(F)_z & \text{if } z \neq x, y, \end{cases}$$

we already have τ on an open subset not containing x, y and the exactness of $(*)$ by the exact sequence (**) below, which was shown in 1.14 (in fact, $\mathscr{M}''_z = \mathscr{M}_z$ and the assertion for $\mathscr{M}''$ is satisfied for $z \neq x, y$).

$$0 \to \mu_*(\mathscr{O}(\mu^* F - \mathfrak{c})) \to \mathscr{O}(F)_C \to \mathscr{M} \to 0. \tag{**}$$

Therefore, we only have to show that such a τ, defined except at x, y, can be extended to the whole surface.

At x, we have the following exact sequence (**)' from (**):

$$0 \to \bigoplus_{\lambda=1}^{r} \mathscr{O}(-\mathfrak{c})_{\mathfrak{p}_\lambda} \xrightarrow{\mu} (\mathscr{O}_C)_x \to \mathscr{M}_x \to 0. \tag{**'}$$

On the other hand, since

$$\bigoplus_{\lambda=1}^{r} \mathscr{O}(-\mathfrak{c}-\mathfrak{d}_x)_{\mathfrak{p}_\lambda} \subset \bigoplus_{\lambda=1}^{r} \mathscr{O}(-\mathfrak{c})_{\mathfrak{p}_\lambda}, \quad (\mathscr{O}(-hx)_C)_x \subset (\mathscr{O}_C)_x,$$

we will finish the proof if we can show

$$\mu(\bigoplus_{\lambda=1}^{r} \mathscr{O}(-\mathfrak{c}-\mathfrak{d}_x)_{\mathfrak{p}_\lambda}) \subset (\mathscr{O}(-hx)_C)_x, \qquad (***)$$

because, if (***) holds, since a similar thing also holds at y, then τ defined except at x, y can be extended by μ, and since

$$\mathscr{M}_x'' = \operatorname{Coker}\left(\bigoplus_{\lambda=1}^{r} \mathscr{O}(-\mathfrak{c}-\mathfrak{d}_x)_{\mathfrak{p}_\lambda} \to (\mathscr{O}(-hx)_C)_x\right),$$

considering that $\dim \mathscr{M}_x < \infty$ and the definition of $\mathfrak{d}_x$, we see $\dim \mathscr{M}_x'' < \infty$ and obtain the following exact sequence (*)':

$$0 \to \bigoplus_{\lambda=1}^{r} \mathscr{O}(-\mathfrak{c}-\mathfrak{d}_x)_{\mathfrak{p}_\lambda} \xrightarrow{\mu} (\mathscr{O}(-hx)_C)_x \to \mathscr{M}_x'' \to 0. \qquad (*)'$$

Now we show (***). Recall that $\mathscr{O}_{\tilde{C},\mathfrak{p}_\lambda} = \mathbb{C}\{t_\lambda\}$, $\mathscr{O}_{\tilde{C}}(-\mathfrak{c})_{\mathfrak{p}_\lambda} = t_\lambda^{c_\lambda}\mathbb{C}\{t_\lambda\}$, $\mathscr{O}_x = \mathbb{C}\{w, z\}$. We denote by f_C the image of $f \in \mathscr{O}_x$ by the restriction map $\mathscr{O}_x \to (\mathscr{O}_C)_x$. Given an element $\xi = \sum_{\lambda=1}^{r} \xi_\lambda(t_\lambda) \in \bigoplus_{\lambda=1}^{r} \mathscr{O}(-\mathfrak{c})_{\mathfrak{p}_\lambda}$ ($\xi_\lambda(t_\lambda) \in t_\lambda^{c_\lambda}\mathbb{C}\{t_\lambda\}$), we infer from 1.5.5 that $f \in \mathscr{O}_x$ satisfying $\mu^*\xi = f_C$ is of the following form:

$$\begin{cases} f = \sum_{j=0}^{m-1} f_j(z) w^{m-1-j} \qquad \left(f_j(z) = \sum_{n=0}^{\infty} f_{jn} z^n\right), \\ \text{moreover, } f_{jn} \text{ is given by the following formula:} \\ f_{jn} = \dfrac{1}{2\pi \mathrm{i}} \sum_{\lambda=1}^{r} \oint \xi_\lambda(t_\lambda) B_j(P_\lambda(t_\lambda), t_\lambda^{m_\lambda}) t_\lambda^{-(n+1)m_\lambda} \sigma_\lambda \, \mathrm{d}t_\lambda. \end{cases}$$

In order to prove (***), it is sufficient to show

$$\xi \in \bigoplus_{\lambda=1}^{r} \mathscr{O}(-\mathfrak{c}-\mathfrak{d}_x)_{\mathfrak{p}_\lambda} \Longrightarrow f \in \mathscr{O}(-hx)_x,$$

that is,

$$\xi_\lambda(t_\lambda) \equiv 0 \ (t_\lambda^{c_\lambda+d_\lambda}) \Longrightarrow f_{jn} = 0 \quad (m-1-j+n \leqq h-1),$$

where $d_\lambda = (h-m+1)^+ m_\lambda$ and, hence, $\mathfrak{d}_x = \sum_{\lambda=1}^r d_\lambda \mathfrak{p}_\lambda$.

We assume that $\xi_\lambda(t_\lambda) \equiv 0 \ (t_\lambda^{c_\lambda+d_\lambda})$ $(1 \leqq \lambda \leqq r)$. Since $B_j(w,z) = w^j + A_1(z)w^{j-1} + \cdots + A_j(z)$, $A_k(z) \in z^k\mathbb{C}\{z\}$, we have $B_j(P_\lambda(t_\lambda), t_\lambda^{m_\lambda}) \equiv 0 \ (t_\lambda^{jm_\lambda})$. Hence

$$\xi_\lambda(t_\lambda) B_j(P_\lambda(t_\lambda), t_\lambda^{m_\lambda}) t_\lambda^{-(n+1)m_\lambda} \sigma_\lambda \equiv 0 \quad (t_\lambda^{d_\lambda + jm_\lambda - (n+1)m_\lambda}).$$

On the other hand, when $m-1-j+n \leqq h-1$,

$$\begin{aligned} d_\lambda + jm_\lambda - (n+1)m_\lambda &\geqq (h-m+1)m_\lambda + jm_\lambda - (n+1)m_\lambda \\ &= \{h-1-(m-1-j+n)\}m_\lambda \geqq 0 \end{aligned}$$

from the definition of d_λ. Hence, for such j, n, we obtain $f_{jn} = 0$ in view of the equation giving f_{jn}. □

6.2 Assume that $\dim H^0(S, \mathcal{O}(F)) = n+1 \geqq 2$ for a line bundle F (we put $\dim|F| = n$ in this case). Let us define the rational map $\Phi_F\colon S \to \mathbb{P}^n$ associated with F. Take a basis $\{\varphi_0, \varphi_1, \ldots, \varphi_n\}$ for $H^0(S, \mathcal{O}(F))$. Choosing a suitable finite open covering $S = \cup_j U_j$ and letting $F = \{f_{jk}\}$, we may assume that $\varphi_\lambda = \{\varphi_{\lambda j}\}$, where $\varphi_{\lambda j}(z)$ is a holomorphic function on U_j satisfying $\varphi_{\lambda j}(z) = f_{jk}(z)\varphi_{\lambda k}(z)$ on $U_j \cap U_k$. Therefore, for $z \in U_j \cap U_k$, as homogeneous coordinates on $\mathbb{P}^n$, we have

$$(\varphi_{0j}(z), \varphi_{1j}(z), \ldots, \varphi_{nj}(z)) = (\varphi_{0k}(z), \varphi_{1k}(z), \ldots, \varphi_{nk}(z))$$

and we can define a rational map $\Phi_F\colon S \to \mathbb{P}^n$ by

$$\Phi_F\colon z \longrightarrow \Phi_F(z) = (\varphi_0(z), \ldots, \varphi_\lambda(z), \ldots, \varphi_n(z)).$$

This is uniquely determined by F modulo projective transformations changing coordinates on $\mathbb{P}^n$.

6.2.1 $z \in S$ is a base point of $|F| \iff \varphi_0(z) = \varphi_1(z) = \cdots = \varphi_n(z) = 0$ (that is, $\Phi_F(z)$ cannot be defined).

6.2.2 $|F|$ has no base points $\Longrightarrow$ Φ_F is a holomorphic map.

Theorem 6.3 *If $F^2 > 0$ and $|mF|$ has no base points for an integer $m > 0$, then $H^1(S, \mathcal{O}(K+F)) = 0$.*

Proof By hypothesis, Φ_{mF} is a holomorphic map. Let $\{\varphi_0, \varphi_1, \dots, \varphi_n\}$ be a basis for $H^0(S, mF)$.

We first show that (a) $\Phi_{mF}(S)$ is a surface. Assume on the contrary that $\Phi_{mF}(S) \subset \mathbb{P}^n$ is a curve. Take a hyperplane L in general position and put $L \cap \Phi_{mF}(S) = \bigcup_{k=1}^{\ell} \mathfrak{p}_k$. We may assume that L and $\Phi_{mF}(S)$ meet normally at $\mathfrak{p}_k$. We let $\sum_{i=0}^{n} a_i \zeta_i = 0$ be the equation of L (where $(\zeta_0, \dots, \zeta_n)$ is a system of homogeneous coordinates on $\mathbb{P}^n$). Putting $\varphi = \sum_{\lambda=0}^{n} a_\lambda \varphi_\lambda$, $D = (\varphi)$, we have

$$\Phi_{mF}^{-1}\Big(\bigcup_{k=1}^{\ell} \mathfrak{p}_k\Big) = \{z \in S \mid \varphi(z) = 0\} = D.$$

If we put $C_k = \Phi_{mF}^{-1}(\mathfrak{p}_k)$, then $D = \bigcup_{k=1}^{\ell} C_k$. Since $\varphi \in H^0(S, \mathcal{O}(mF))$, we see $D \in |mF|$. Similarly, taking a hyperplane L': $\sum_{\lambda=0}^{n} a'_\lambda \zeta_\lambda = 0$ in general position with $L \cap L' \cap \Phi_{mF}(S) = \emptyset$, and putting $\varphi' = \sum_{\lambda=0}^{n} a'_\lambda \varphi_\lambda$ and $D' = (\varphi')$, we get $D' = \Phi_{mF}^{-1}(\bigcup \mathfrak{p}'_k)$, where $L' \cap \Phi_{mF}(S) = \bigcup_{k=1}^{\ell} \mathfrak{p}'_k$, and $D' \cap D = \emptyset$. Obviously $D' \in |mF|$. Therefore, $m^2 F^2 = (mF)^2 = D \cdot D' = 0$, which contradicts the assumption that $F^2 > 0$. We have shown that $\Phi_{mF}(S)$ is a surface (because it is immediate that $\dim \Phi_{mF}(S) > 0$).

Since $H^1(S, \mathcal{O}(K + F)) \cong \mathbb{H}^{2,1}(F)$, where $\mathbb{H}^{2,1}(F)$ denotes the space of harmonic (2, 1)-forms with coefficients in F, we next show that (b) $\mathbb{H}^{2,1}(F) = 0$, which will complete the proof. We let $S = \bigcup U_j$ be a sufficiently fine open covering and $F = \{f_{jk}(z)\}$. Any $u \in \mathbb{H}^{2,1}(F)$ can be written as $u = u_j = \sum_{\alpha=1}^{2} u_{j12\bar{\alpha}} dz^1 \wedge dz^2 \wedge d\bar{z}^\alpha$ on U_j and satisfies $u_j = f_{jk} u_k$ on $U_j \cap U_k$. Moreover, we have $\Box u = 0$ by definition, where $\Box$ denotes the complex Laplace–Beltrami operator. We shall show $u = 0$. Letting $\varphi_\lambda = \{\varphi_{\lambda j}(z)\}$ $(0 \leqq \lambda \leqq n)$, we have $\varphi_{\lambda j}(z) = f_{jk}^m \cdot \varphi_{\lambda k}(z)$ on $U_j \cap U_k$. We define things as follows:

6.3.1 $$a_j(z) = \left(\sum_{\lambda=0}^{n} |\varphi_{\lambda j}(z)|^2\right)^{\frac{1}{m}} > 0 \quad (\text{on } U_j).$$

6.3.2 $\gamma = \frac{i}{2\pi} \partial\bar{\partial} \log a_j(z)$ (on each U_j).

Then γ is a closed real (1, 1)-form on S (since $a_j(z) = |f_{jk}(z)|^2 a_k(z)$, we get $\partial\bar{\partial} a_j = \partial\bar{\partial} a_k$ on $U_j \cap U_k$). We express γ as

6.3.3 $$\gamma = \frac{i}{2\pi} \sum_{\alpha,\beta=1}^{2} \gamma_{\alpha\bar{\beta}}\, dz^\alpha \wedge d\bar{z}^\beta.$$

Using a Kähler metric $\sum_{\alpha,\beta=1}^{2} g_{\alpha\bar{\beta}} dz^\alpha d\bar{z}^\beta$ on S, we define $\gamma^{\bar{\beta}\alpha}$ by the rule $\gamma_{\lambda\bar{\nu}} = \sum_{\alpha=1}^{2}\sum_{\beta=1}^{2} g_{\lambda\bar{\beta}} g_{\alpha\bar{\nu}} \gamma^{\bar{\beta}\alpha}$. Then the following holds (Kodaira [4]):

$$\int_S \sum_{\alpha,\beta=1}^{2} \gamma^{\bar{\beta}\alpha} \frac{1}{a_j} u_{j12\bar{\alpha}} \overline{u_{j12\bar{\beta}}}\, dz^1 \wedge dz^2 \wedge \overline{dz^1} \wedge \overline{dz^2} \leqq 0. \tag{*}$$

Now, if we let $\mathbb{P}^n = \bigcup_{\lambda=0}^{n} \mathscr{U}_\lambda$ ($\mathscr{U}_\lambda = \{\zeta \in \mathbb{P}^n \mid \zeta_\lambda \neq 0\}$) and put $A_\lambda = \dfrac{\sum_{\nu=0}^{n} |\zeta_\nu|^2}{|\zeta_\lambda|^2} = 1 + \sum_{\nu\neq\lambda} \left|\dfrac{\zeta_\nu}{\zeta_\lambda}\right|^2$, then $\omega = \dfrac{i}{2\pi}\partial\bar{\partial}\log A_\lambda$ is a Kähler form on $\mathbb{P}^n$. Recall that $\Phi_{mF}: S \to \mathbb{P}^n$ is a holomorphic map. We have $a_j^m = \Phi_{mF}^*(A_\lambda)$ and, hence, $m\gamma = \Phi_{mF}^*(\omega)$. Since ω is positive definite, we see that γ is positive semi-definite. On the other hand, since $\Phi_{mF}(S)$ is a surface, there exists an analytic subspace $N \not\subseteq S$ such that γ is positive definite on $S \setminus N$ and Φ_{mF} is a local biholomorphic map there. We have, however, $\int_{S\setminus N} \gamma u \bar{u} dz \cdots \leqq 0$ by $(*)$. Since $\gamma > 0$ on $S \setminus N$, we have $u = 0$ on $S \setminus N$. Since $\Box u = 0$, u is C^∞ on S. Therefore, we have $u = 0$ on S. □

Remark 6.3.1 In Mumford [6], we can find (1) an algebraic proof of Theorem 6.3 in characteristic 0 and (2) a counterexample in characteristic $p \neq 0$.

2.7 Composition Series

Assumption 7.0 *Throughout, we let S denote a non-singular minimal algebraic surface of general type, that is,*

$$\begin{cases} \text{(a) } S \text{ contains no exceptional curves of the first kind,} \\ \text{(b) } P_2 = \dim|2K| + 1 > 0, \\ \text{(c) } c_1^2 = K^2 > 0. \end{cases}$$

Notation 7.1 We call a finite sum $\xi = \sum r_i C_i$ ($r_i \in \mathbb{Q}$) a *divisorial cycle*. For divisorial cycles $\xi = \sum r_i C_i$, $\eta = \sum s_j C_j$, we put

$$\xi \cdot \eta = \sum\sum r_i s_j C_i \cdot C_j.$$

In the following, we indicate by the symbol $\sim$ homology with respect to rational coefficients.

Lemma 7.2 *For a divisorial cycle* $\zeta = \sum r_i C_i$,

$$\zeta \not\sim 0 \text{ and } K \cdot \zeta = 0 \Longrightarrow \zeta^2 < 0.$$

Proof It follows from de Rham's theorem that $H^2(S, \mathbb{R}) \cong Z^2/\mathrm{d}A^1$, where $Z^2 = \{\varphi \mid \varphi \text{ is a 2-form}, \mathrm{d}\varphi = 0\}$ and $A^1 = \{\eta \mid \eta \text{ is a 1-form}\}$. For $\varphi, \psi \in Z^2$, we put $(\varphi, \psi) = \int_S \varphi \wedge \psi$. Then this is a symmetric bilinear form on $Z^2/\mathrm{d}A^1$ (because $(\varphi, \mathrm{d}\eta) = \int_S \varphi \wedge \mathrm{d}\eta = \int_S \mathrm{d}(\varphi \wedge \eta) = 0$). If we choose a suitable basis $\{\beta_1, \ldots, \beta_i, \ldots, \beta_b\}$ (b is the second Betti number of S), then we have

$$((\beta_i, \beta_k)) = \begin{pmatrix} +1 & & & & & \\ & \ddots & & & O & \\ & & +1 & & & \\ & & & -1 & & \\ & O & & & \ddots & \\ & & & & & -1 \end{pmatrix}$$

in which the number of $+1$ is $2p_g + 1$ by Hodge's index theorem. Here p_g is the number of linearly independent holomorphic 2-forms and we take such forms $\psi_1, \ldots, \psi_{p_g}$ so that $\int_S \psi_i \wedge \bar{\psi}_k = \delta_{ik}$. If we put $\omega_i = \frac{1}{\sqrt{2}}(\psi_i + \bar{\psi}_i)$ and $\omega_{p_g+i} = \frac{1}{\sqrt{-2}}(\psi_i - \bar{\psi}_i)$ $(1 \leqq i \leqq p_g)$, then $(\omega_i, \omega_k) = \delta_{ik}$ and $\omega_1, \ldots, \omega_{2p_g}$ are linearly independent. We take new $\beta_1, \ldots, \beta_{b-2p_g}$ and complete a basis for $Z^2/\mathrm{d}A^1$ satisfying

$$\begin{cases} Z^2/\mathrm{d}A^1 = \{\omega_1, \ldots, \omega_{2p_g}\} \bigoplus \{\beta_1, \ldots, \beta_{b-2p_g}\}, \\ ((\beta_i, \beta_k)) = \begin{pmatrix} +1 & & & \\ & -1 & & \\ & & \ddots & \\ & & & -1 \end{pmatrix}. \end{cases}$$

In general, if X is a divisor, then we can take a 2-form $\xi \in Z^2$ corresponding to the Chern class $c([X])$ (de Rham's theorem). Similarly, we take $\eta \in Z^2$ for a divisor Y. Recall that we have $X \cdot Y = \int_S \xi \wedge \eta$ and $-\int_X \psi = \int_S \xi \wedge \psi$ for any 2-form ψ. Noting that $\int_S \xi \wedge \omega_i = 0$ (by $\int_X \omega_i = 0$), we take a closed 2-form β corresponding to K. Since $K^2 > 0$, we have $\beta^2 > 0$ and we can assume that $\beta_1 = \beta$.

Now, from the assumption that $K \cdot \zeta = 0$, we get $\zeta \sim \sum_{i \geqq 2} a_i \beta_i$. Furthermore, some a_j cannot be zero, since $\zeta \not\sim 0$. Hence $\zeta^2 = \sum_{i \geqq 2} a_i^2 \beta_i^2 < 0$. □

Remark 7.2.1 An algebraic proof of 7.2 can be found in Zariski [9].

Remark 7.2.2 If $\zeta = \sum r_i C_i > 0$ (i.e., $r_i > 0$), then $\zeta \not\sim 0$.

Lemma 7.3 *For any irreducible curve C, $KC \geqq 0$ holds. In particular,*

$$KC = 0 \Longleftrightarrow \begin{cases} \text{(a) } C \text{ is a non-singular rational curve, i.e., } \pi(C) = 0, \\ \text{(b) } C^2 = -2. \end{cases}$$

Proof Since $\dim|2K| = P_2 - 1 \geqq 0$, we have $|2K| \ni \exists D > 0$. Write $D = kC + \sum_{i=1}^{\ell} k_i C_i$ $(k, k_i \geqq 0, C \neq C_i)$. Then $2K \cdot C = D \cdot C = kC^2 + \sum_{i=1}^{\ell} k_i C \cdot C_i$. Since $k, k_i \geqq 0$ and $C \cdot C_i \geqq 0$, if $KC < 0$, then $k > 0$ and $C^2 < 0$. Then by the formula $2\pi(C) - 2 = KC + C^2$, we must have $KC = C^2 = -1$ and $\pi(C) = 0$. Hence C is an exceptional curve of the first kind and this contradicts Assumption 7.0. Therefore, $KC \geqq 0$.

As to the latter part: ($\Rightarrow$) Assume that $KC = 0$. We have $C \not\sim 0$, because, in the situation $C \subset S \hookrightarrow \mathbb{P}^n$, we have $C \cdot L > 0$ for a hyperplane L in $\mathbb{P}^n$. Then $2\pi(C) - 2 = KC + C^2 = C^2 < 0$ and it follows that $\pi(C) = 0$ and $C^2 = -2$. ($\Leftarrow$) Clear from $2\pi(C) - 2 = KC + C^2$. □

Remark 7.3.1 This can be found in Mumford [5].

Remark 7.4 If $KC = 0$, then the restriction K_C of K to C is the trivial line bundle. Therefore, so is mK_C and any $\varphi \in H^0(S, \mathcal{O}(mK))$ is constant on C. Then $\Phi_{mK}(C)$ is a point.

Because such curves are obstacles to our purpose, we shall study their properties further.

Theorem 7.5 *The number of irreducible curves C with $KC = 0$ is less than the second Betti number b_2 of S.*

Proof Let $KC_i = 0$ $(i = 1, 2, \ldots, \ell)$ $(C_i \neq C_k$ if $i \neq k)$. It suffices to show that the C_i's are homologically independent. Let $\sum_{i=1}^{\ell} r_i C_i \sim 0$, where $r_i \in \mathbb{Q}$, $r_1, \ldots, r_p \geqq 0$, $r_{p+1}, \ldots, r_\ell \leqq 0$. If we put $D = \sum_{i=1}^{p} r_i C_i$, then $D \sim \sum_{j=p+1}^{\ell} (-r_j) C_j = \sum_{j=p+1}^{\ell} |r_j| C_j$. Since $K \cdot D = 0$, we infer from 7.2 that $D^2 \leqq 0$. On the other hand,

$$D^2 = \sum_{1 \leqq i \leqq p} r_i C_i \cdot \sum_{p+1 \leqq j \leqq \ell} |r_j| C_j = \sum_{\substack{1 \leqq i \leqq p, \\ p+1 \leqq j \leqq \ell}} r_i |r_j| C_i \cdot C_j \geqq 0.$$

Therefore, $D^2 = 0$ and, by 7.2 again, $D \sim 0$. Considering $S \subset \mathbb{P}^n$, if L is a hyperplane in $\mathbb{P}^n$, then $0 = D \cdot L = \sum_{i=1}^{p} r_i L \cdot C_i$. Since $L \cdot C_i > 0$, we obtain $r_i = 0$. Similarly, $r_j = 0$ $(p + 1 \leqq j \leqq \ell)$. □

Remark 7.5.1 If $C_1, \ldots, C_\ell$ are curves as in 7.5, then $K \not\sim \sum r_i C_i$, that is, K and $\{C_i\}$ are homologically independent. In fact, if $K \sim \sum r_i C_i$, then $K^2 = \sum r_i (C_i \cdot K) = 0$, which contradicts $K^2 > 0$.

Notation 7.6 Let $E_1, E_2, \ldots, E_b$ $(b < b_2)$ be all the irreducible curves on S satisfying $KE_i = 0$. We put $\mathscr{E} = E_1 + E_2 + \cdots + E_b$. We write it as a sum $\mathscr{E} = \mathscr{E}_1 + \cdots + \mathscr{E}_\lambda + \cdots + \mathscr{E}_\rho$ of connected components $\mathscr{E}_\lambda$.

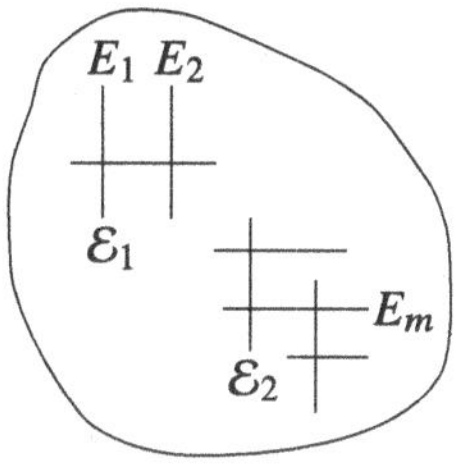

Assumption 7.7 *In the following, we denote by e a positive integer such that* $\dim|eK| \geqq 0$. *Take* $D \in |eK|$ *and write* $D = C_1 + C_2 + \cdots + C_i + \cdots + C_n$ *(C_i: irreducible curve, but we do not assume $C_i \neq C_k$ when $i \neq k$). If we do not mind the order, the set $\{C_i\}$ is uniquely determined. However, we take the order into consideration and call the presentation* $D = C_1 + C_2 + \cdots + C_i + \cdots + C_n = \sum_{i=1}^n C_i$ *a* composition series *of D. Put* $D_i = C_1 + \cdots + C_i$. *We consider the following two conditions:*

(α) $K \cdot C_1 \geqq 1,\ D_{i-1} \cdot C_i \geqq 1 \quad (i = 2, 3, \ldots, n).$
(β) $K \cdot C_1 \geqq 1,\ D_{i-1} \cdot C_i \geqq 0,\ KC_i + D_{i-1}C_i \geqq 1 \quad (i = 2, 3, \ldots, n).$

We study whether it is possible to find a composition series satisfying (α) or (β) (7.9 and 7.12). It is clear that (α) implies (β) (7.3). Until 7.10, we denote by X, Y divisors.

Lemma 7.8 $D = X + Y,\ X > 0,\ Y > 0 \implies X \cdot Y \geqq 1.$

Proof If we put

$$r = \frac{K \cdot X}{K^2}, \quad \xi = X - rK, \quad s = \frac{K \cdot Y}{K^2}, \quad \eta = Y - sY,$$

then $X = rK + \xi$, $K \cdot \xi = 0$, $Y = sK + \eta$, $K \cdot \eta = 0$. Since ξ, η are divisorial cycles and $K \cdot X \geqq 0$, $K \cdot Y \geqq 0$, we have $r, s \geqq 0$. Since $D \in |eK|$, we have $D \sim eK$ and $D = X + Y = (r+s)K + \xi + \eta$. Hence $eK^2 = K \cdot D = (r+s)K^2$ and we get $e = r + s$, $\xi + \eta \sim 0$. We have $X \cdot Y = rsK^2 + \xi\eta = rsK^2 - \xi^2$. We infer from 7.2 that $\xi^2 \leqq 0$. Hence $X \cdot Y \geqq 0$ because $r, s \geqq 0$. If $X \cdot Y = 0$, then $\xi^2 = 0$ and $rs = 0$. Hence $\xi \sim 0$ and, similarly, $\eta \sim 0$ (7.2). Then we have $X \sim rK$, $Y \sim sK$. Since $rs = 0$, however, we have either $X \sim 0$ or $Y \sim 0$. This contradicts that $X > 0$, $Y > 0$ (7.2.2). Consequently, $X \cdot Y > 0$. □

Lemma 7.9 *There exists a composition series* $D = \sum_{i=1}^n C_i$ *of D satisfying the condition (α).*

Proof Since $D \in |eK|$, $K \cdot D = eK^2 \geqq e > 0$. Writing $D = \sum_{i=1}^n \Theta_i$ (Θ_i: irreducible), since $K \cdot D = \sum_{i=1}^n K \cdot \Theta_i > 0$, there exists at least one Θ_i satisfying

$K \cdot \Theta_i > 0$. Putting $C_1 = \Theta_i$, we have $K \cdot C_1 \geqq 1$. We write $D = C_1 + Y$. Then we have $Y \geqq 0$. If $Y = 0$, then we are done. If $Y > 0$, then $C_1 \cdot Y \geqq 1$ by 7.8. Hence for $Y = \sum_{j \neq i} \Theta_j$, we can find Θ_j satisfying $C_1 \Theta_j > 0$. We put $C_2 = \Theta_j$. Then $C_1 \cdot C_2 = D_1 \cdot C_2 \geqq 1$. We argue by induction. Suppose that we have chosen $C_1 = \Theta_{\nu_1}, C_2 = \Theta_{\nu_2}, \ldots, C_{i-1} = \Theta_{i-1}$ so that $D_{j-1} C_j \geqq 1$ $(j = 2, 3, \ldots, i-1)$, where $D_{j-1} = C_1 + C_2 + \cdots + C_{j-1}$. Writing $D = D_{i-1} + Y'$, if $Y' = 0$ then we are done; if $Y' > 0$, then $D_{i-1} \cdot Y \geqq 1$ by 7.8 and we can find $\Theta_\lambda < Y'$ satisfying $D_{i-1} \cdot \Theta_\lambda \geqq 1$. Hence, letting $C_i = \Theta_\lambda$, we obtain $D_{i-1} \cdot C_i \geqq 1$. □

Remark 7.9.1 From the proof, if we are given $\Theta < D$ such that $K \cdot \Theta \geqq 1$, then we can put $C_1 = \Theta$ in finding a composition series $D = \sum_{i=1}^n C_i$ satisfying (α).

Lemma 7.10 *Assume that $K \cdot E_p = K \cdot E_q = E_p \cdot E_q = 0$ (7.6). For $D \in |eK|$, if $D = X + Y + E_p + E_q$, $X > 0$, $Y > 0$, $K \cdot X > 0$ and $K \cdot Y > 0$, then $X \cdot Y \geqq 0$.*

Proof We put

$$r = \frac{K \cdot X}{K^2}, \quad r_1 = \frac{E_p \cdot X}{E_p^2} = -\frac{E_p \cdot X}{2}, \quad r_2 = \frac{E_q \cdot X}{E_q^2} = -\frac{E_q \cdot X}{2}$$

and $\xi = X - rK - r_1 E_p - r_2 E_q$ (recall that we have $E_p^2 = E_q^2 = -2$ by 7.3). Then $K \cdot \xi = E_p \cdot \xi = E_q \cdot \xi = 0$ by hypothesis. Similarly, if we put

$$s = \frac{K \cdot Y}{K^2}, \quad s_1 = -\frac{E_p \cdot Y}{2}, \quad s_2 = -\frac{E_q \cdot Y}{2}$$

and $\eta = Y - sK - s_1 E_p - s_2 E_q$, then $K \cdot \eta = E_p \cdot \eta = E_q \cdot \eta = 0$.

On the other hand, we have $r + s = e$ and $r_1 + s_1 + 1 = r_2 + s_2 + 1 = 0$ from $K \cdot (X + Y) = K \cdot D - K \cdot E_p - K \cdot E_q = eK^2$, $E_p \cdot (X + Y) = E_p \cdot D - E_p^2 - E_p \cdot E_q = 2$ and $E_q \cdot (X + Y) = 2$. Hence

$$\begin{aligned} D = X + Y + E_p + E_q &= (r + s)K + (r_1 + s_1 + 1)E_p + (r_2 + s_2 + 1)E_q + \xi + \eta \\ &= eK + \xi + \eta \end{aligned}$$

and it follows that $\xi + \eta \sim 0$.

We also have $X \cdot Y = rsK^2 + r_1 s_1 E_p^2 + r_2 s_2 E_q^2 + \xi \cdot \eta$. Since $rK^2 = sK^2 = K \cdot X \geqq 1$, we obtain $rsK^2 \geqq \frac{1}{2}(r+s) = \frac{e}{2}$ from $rsK^2 \geqq r, s$. Moreover, $r_1 s_1 E_p^2 = -2r_1 s_1 = 2s_1(s_1+1) = 2(s_1+\frac{1}{2})^2 - \frac{1}{2} \geqq -\frac{1}{2}$ and, similarly, $r_2 s_2 E_q^2 \geqq -\frac{1}{2}$. Finally, we have $\xi \cdot \eta = -\xi^2 \geqq 0$ (because $\xi + \eta \sim 0$ and $\xi^2 \leqq 0$ from $K \cdot \xi = 0$ by 7.2). Therefore, $X \cdot Y \geqq \frac{e}{2} - \frac{1}{2} - \frac{1}{2} - \xi^2 \geqq \frac{e}{2} - 1 > -1$, that is, $X \cdot Y \geqq 0$. □

Remark 7.11 In the expression $\mathscr{E} = \mathscr{E}_1 + \cdots + \mathscr{E}_\lambda + \cdots + \mathscr{E}_\rho$ as in Notation 7.6, if $D \in |eK|$ meets $\mathscr{E}_\lambda$, then $\mathscr{E}_\lambda < D$. In fact, write $\mathscr{E}_\lambda = \sum E_i$ $(K \cdot E_i = 0$, $\pi(E_i) = 0$, $E_i^2 = -2)$. Then it follows from $D \sim eK$ that $D \cdot E_i = eK \cdot E_i = 0$. Since $D > 0$, this implies that $E_i < D$ when $D \cap E_i \neq \emptyset$. Since $\mathscr{E}_\lambda$ is connected, we have $\mathscr{E}_\lambda < D$.

Lemma 7.12 *If D meets $\mathscr{E}_\lambda$ and $\mathscr{E}_\nu$ $(\lambda \neq \nu)$ (this is equivalent to saying that $\mathscr{E}_\lambda < D$ and $\mathscr{E}_\nu < D$ by 7.11), then there exists a composition series $D = \sum_{i=1}^{n} C_i$ satisfying (β) with $C_n < \mathscr{E}_\nu$ and $C_{n-1} < \mathscr{E}_\lambda$.*

Proof We take $E_p < \mathscr{E}_\lambda$ and $E_q < \mathscr{E}_\nu$. Since $K \cdot D = eK^2 > 0$, we can find an irreducible component $\Theta < D$ satisfying $K \cdot \Theta > 0$ and put $C_1 = \Theta$. Therefore, $K \cdot C_1 \geqq 1$. Next, assume that we have chosen components $C_1, C_2, \ldots, C_i$ $(i \geqq 2)$ of D so that

$$\begin{cases} (P_i)\colon D = C_1 + C_2 + \cdots + C_{j-1} + C_j + \cdots + C_{i-1} + X_i + E_p + E_q, \ X_i \geqq 0, \\ (\beta_i)\colon D_{j-1} \cdot C_j \geqq 0, \ K \cdot C_j + D_{j-1} \cdot C_j \geqq 1 \ (1 \leqq j \leqq i-1), \end{cases}$$

where $D_{j-1} = C_1 + \cdots + C_{j-1}$. When $i = 2$, these merely say $K \cdot C_1 \geqq 1$. In general, we have $K \cdot X_i \geqq 0$ by 7.3. We shall show that there exists $i_0 \geqq i$ satisfying (P_{i_0}), (β_{i_0}) and $K \cdot X_{i_0} = 0$. In fact, if $K \cdot X_i > 0$, then writing $D = D_{i-1} + X_i + E_p + E_q$, $K \cdot D_{i-1} \geqq K \cdot C_1 \geqq 1$ (7.3). Then we infer from 7.10 that $D_{i-1} \cdot X_i \geqq 0$. There are the following two possibilities:

$$\begin{cases} \text{(a) There exists a component } C < X \text{ with } D_{i-1} \cdot C > 0. \\ \text{(b) } D_{i-1} \cdot C = 0 \text{ for all } C < X_i. \end{cases}$$

If (a) is the case, then, putting $C_i = C$, $X_{i+1} = X_i - C$, $D_{i-1} \cdot C_i \geqq 1$. If (b) is the case, then, by $K \cdot X_i > 0$, there exists a component $C < X_i$ satisfying $K \cdot C > 0$. We put $C_i = C$. Then, by (b), we have $D_{i-1} \cdot C_i = 0$ and $K \cdot C_i + D_{i-1} \cdot C_i \geqq 1$. Therefore, in either case, we can choose C_i so that we have (P_{i+1}), (β_{i+1}), $K \cdot X_{i+1} \geqq 0$. Moreover, we obviously have $K \cdot X_i \geqq K \cdot X_{i+1} (\geqq 0)$. Since we can continue the procedure until we arrive at i_0 with $K \cdot X_{i_0} = 0$, we conclude that there exists $i_0 \geqq i (\geqq 2)$ satisfying

$$\begin{cases} (P_{i_0})\colon D = C_1 + C_2 + \cdots + C_{i_0-1} + X_{i_0} + E_p + E_q, \ X_{i_0} \geqq 0, \\ (\beta_{i_0})\colon D_{j-1} \cdot C_j \geqq 0, \ K \cdot C_j + D_{j-1} \cdot C_j \geqq 1 \ (1 \leqq j \leqq i_0 - 1), \\ \text{and} \quad K \cdot X_{i_0} = 0. \end{cases}$$

As in the proof of 7.9, using 7.8, we can also choose components $C_{i_0+1}, \ldots, C_n$ of $X_{i_0} + E_p + E_q$ so that

$$\begin{cases} D = C_1 + \cdots + C_{i_0} + C_{i_0+1} + \cdots + C_\ell + \cdots + C_h + \cdots + C_n \ (C_\ell = E_p, C_h = E_q), \\ (\alpha)\colon \ D_{j-1} \cdot C_j = (C_1 + \cdots + C_{j-1}) \cdot C_j \geqq 1 \quad (2 \leqq j \leqq n). \end{cases}$$

We shall show that we can change the numbering of the C_i's so that

$$D_{j-1} \cdot C_j \geqq 1 \quad (2 \leqq j \leqq n), \ C_{n-1} < \mathscr{E}_\lambda, C_n < \mathscr{E}_\nu$$

if necessary. We remark that, if $C_i \cdot C_{i+1} = 0$, we can exchange them to get (α), because putting (α_i): $(C_1 + \cdots + C_{i-1}) \cdot C_i \geqq 1$, we infer from $C_i \cdot C_{i+1} = 0$ and (α_i)

that $(C_1 + \cdots + C_{i-1} + C_{i+1}) \cdot C_i \geqq 1$ and, from $C_i \cdot C_{i+1} = 0$, (α_{i+1}) and (α_{i+2}) that $(C_1 + \cdots + C_{i-1}) \cdot C_{i+1} \geqq 1$ and $(C_1 + \cdots + C_{i-1} + C_{i+1} + C_i) \cdot C_{i+2} \geqq 1$.

Note that every C_i $(i > i_0)$ is a component of $\mathscr{E}$, because $K \cdot X_{i_0} = 0$. Since two curves in different connected components do not meet, we can apply the argument just given and gather curves in the same connected component of $\mathscr{E}$, assigning a consecutive number to them. Therefore, after C_{i_0}, we can choose the numbering so that components of $\mathscr{E}_\lambda$, $\mathscr{E}_\nu$ are separated as

$$D = C_1 + \cdots + C_{i_0} + \underbrace{C_{i_0+1} + \cdots + E_p + \cdots + C_s}_{<\mathscr{E}_\lambda} + \underbrace{\cdots + E_q + \cdots + C_n}_{<\mathscr{E}_\nu}.$$

(In D, there are no components of $\mathscr{E}_\lambda$, $\mathscr{E}_\nu$ other than indicated.) Furthermore, because we can move C_s to the position just before C_n similarly, it is possible to set $C_s = C_{n-1}$. □

Lemma 7.13 *Given $\mathscr{E}_\lambda$ meeting $D \in |eK|$, there exists a composition series $D = \sum_{i=1}^{n} C_i$ of D which satisfies $C_n < \mathscr{E}_\lambda$ and the condition (β).*

Proof Similar to the proof of 7.12. □

2.8 Conclusions

Notation 8.0 We have chosen and fixed a positive integer e satisfying $\dim |eK| \geqq 0$ (7.7). Taking a member $D \in |eK|$ and a composition series $D = \sum_{i=1}^{n} C_i$ of D, we choose and fix an integer m with $m > e$ and put

$$F_i = mK - [Z_i] \qquad (\text{where } Z_i = C_i + \cdots + C_n).$$

(We shall employ this notation to the end.) Recall that $D_i = C_1 + \cdots + C_i$ (7.7).

8.1 $F_{i+1} \cdot C_i - C_i^2 - K \cdot C_i = (m - e - 1)K \cdot C_i + D_{i-1} \cdot C_i.$

Proof Clear from $F_{i+1} - [C_i] - K = (m-1)K - [C_i + Z_{i+1}] \sim (m - 1 - e)K + [D - Z_i] = (m - 1 - e)K + [D_{i-1}]$. □

8.2 Take integers $h, k \geqq 0$. For $x, y \in D$ (where $D \in |eK|$), if the multiplicities of D at x, y are $\geqq h$, $\geqq k$, respectively, then we put $\Xi_i = \mathscr{O}(mK - Z_i - hx - ky)$. (In what follows, for Ξ_i, such conditions are assumed to be satisfied for preassigned h, k, x and y.)

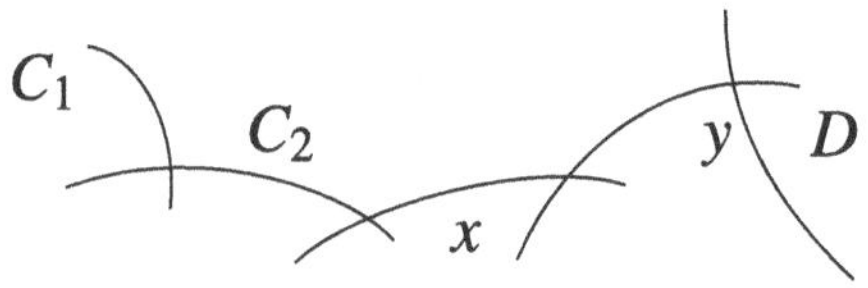

In particular,

$$\Xi_1 = \mathcal{O}(mK - D - hx - ky) = \mathcal{O}(mK - D) \cong \mathcal{O}((m-e)K).$$

8.2.1 $\Xi_1 \subset \Xi_2 \subset \cdots \subset \Xi_i \subset \cdots \subset \Xi_{n+1}$, where $\Xi_{n+1} = \mathcal{O}(mK - hx - ky)$.

Assumption 8.3 *For h, k as above, we let h_i, k_i be integers determined by*

$$\Xi_{i+1}/\Xi_i \cong \mathcal{O}(F_{i+1} - h_i x - k_i y)_{C_i}.$$

8.3.1 If $h = k = 1$, $x, y \in C_i$ and $x, y \notin Z_{i+1}$, then $h_i = k_i = 1$.

This can be seen as follows. In this case, we have $\Xi_{i+1} = \mathcal{O}(mK - Z_{i+1} - x - y) \cong \mathcal{O}(F_{i+1} - x - y)$, $\Xi_i = \mathcal{O}(mK - Z_{i+1} - C_i) \cong \mathcal{O}(F_{i+1} - C_i - x - y)$ and, hence, $\Xi_{i+1}/\Xi_i = \mathcal{O}(F_{i+1} - x - y)/\mathcal{O}(F_{i+1} - x - y - C_i) \cong \mathcal{O}(F_{i+1} - x - y)_{C_i}$.

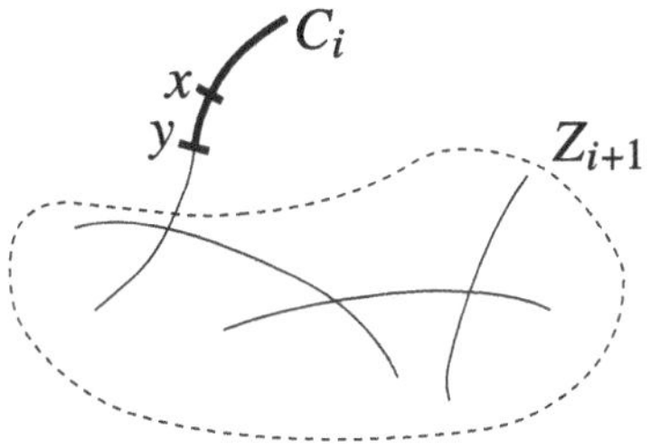

Lemma 8.4 *If* $(m - e - 1)K \cdot C_i + D_{i-1} \cdot C_i > \frac{1}{4}(h_i + 1)^2 + \frac{1}{4}(k_i + 1)^2$, *then* $\dim H^1(S, \mathcal{O}((m-e)K)) \geqq \dim H^1(S, \Xi_i)$ $(2 \leqq i \leqq n+1)$.

Proof For $\alpha \geqq 0$, we have $\frac{1}{4}(h_i+1)^2 \geqq (h_i - \alpha + 1)^+\alpha$. (Because: it is clear when $h_i - \alpha + 1 < 0$. When $h_i - \alpha + 1 \geqq 0$, $(h_i - \alpha + 1)^+\alpha = -(\alpha - \frac{h_i+1}{2})^2 + \frac{1}{4}(h_i+1)^2 \leqq \frac{1}{4}(h_i+1)^2$.) We also have a similar inequality for k_i. Then by hypothesis and 8.1 we infer from 6.1 that $H^1(C_i, \mathcal{O}(F_{i+1} - h_i x - k_i y)_{C_i}) = 0$. From the exact sequence

$$0 \to \Xi_i \to \Xi_{i+1} \to \mathcal{O}(F_{i+1} - h_i x - k_i y)_{C_i} \to 0,$$

we have the exact sequence

$$\cdots \to H^1(S, \Xi_i) \to H^1(S, \Xi_{i+1}) \to 0$$

and, hence, $\dim H^1(S, \Xi_i) \geqq \dim H^1(S, \Xi_{i+1})$. On the other hand, we know $\Xi_1 = \mathcal{O}((m-e)K)$ (8.2). □

Lemma 8.5 (Zariski [9]) *There is an integer m_0 such that, for any $m \geqq m_0$,* $\dim H^1(S, \mathcal{O}((m-e)K)) = \dim H^1(S, \mathcal{O}(mK))$.

Proof For $D \in |eK|$, we take a composition series $D = \sum_{i=1}^n C_i$ of D satisfying condition (α) (7.7 and 7.9). If $m \geqq e+2$, then, since $m - e - 1 \geqq 1$, from 7.3 and (α), we get $(m-e-1)K\cdot C_i+D_{i-1}\cdot C_i \geqq 1$. Putting $h=k=0$, if we put $\Xi_i = \mathcal{O}(mK - Z_i)$, then $h_i = k_i = 0$ (here, one can take any $x, y \in D$). Therefore, the assumption of 8.4 is satisfied and we obtain $\dim H^1(S, \mathcal{O}((m-e)K)) \geqq \dim H^1(S, \mathcal{O}(mK))$, If we put $f(m) = \dim H^1(S, \mathcal{O}(mK))$, then

$$f(m) \geqq f(m+e) \geqq f(m+2e) \geqq \cdots \geqq f(m+ke) \geqq \cdots.$$

Since, however, $f(m)$ is a non-negative integer, there exists an integer ℓ_0 such that $f(m+\ell e)$ is constant for any $\ell \geqq \ell_0$. Hence, we can find m_0 as desired. □

Theorem 8.6 *Assume that $P_e \geqq 2$, $eK^2 \geqq 2$, $m \geqq e+2$ and $m \geqq m_0$. Then the following hold*:

(a) *For any $x \in S$, the following sequence is exact:*

$$0 \to H^0(S, \mathcal{O}(mK - x)) \to H^0(S, \mathcal{O}(mK)) \to \mathbb{C} \to 0.$$

(b) *The m-canonical system $|mK|$ has no base points and Φ_{mK} is a holomorphic map.*

Proof Firstly, recall that we have the following exact sequence for any $x \in S$ (5.0.1):

$$0 \to \mathcal{O}(mK - x) \to \mathcal{O}(mK) \to \mathbb{C}_x \to 0.$$

Hence, we have the exact sequence

$$\begin{aligned} 0 &\to H^0(S, \mathcal{O}(mK - x)) \to H^0(S, \mathcal{O}(mK)) \xrightarrow{\tau} \mathbb{C} \\ &\to H^1(S, \mathcal{O}(mK - x)) \xrightarrow{\lambda} H^1(S, \mathcal{O}(mK)) \to 0, \end{aligned} \tag{*}$$

where $\tau\colon \varphi \to \varphi(x)$ for $\varphi \in H^0(S, \mathcal{O}(mK))$. We shall show that τ is surjective (hence (a)). Then (b) follows immediately.

Since $\dim |eK| \geqq 1$, we can find $D \in |eK|$ satisfying $x \in D$. We consider two cases (I) $x \notin \mathscr{E}$, and (II) $x \in \mathscr{E}$, separately.

(I) $x \notin \mathscr{E}$: By 7.9, we can take a composition series $D = \sum_{i=1}^n C_i$ of D satisfying

$$(\alpha) \quad KC_1 \geqq 1,\ D_{i-1}\cdot C_i \geqq 1 \quad (i \geqq 2).$$

If there exists a component Θ of D satisfying $x \notin \Theta$ and $K\cdot\Theta \geqq 1$, then we can assume that $C_1 = \Theta$ (7.9.1). We assume that $x \in C_\ell, x \notin C_{\ell+1}+\cdots+C_n$. Then we have $K\cdot C_\ell \geqq 1+\delta_{\ell 1}$. (This can be seen as follows. Since $C_\ell \not\subset \mathscr{E}$, we have $K\cdot C_\ell \geqq 1$. If $\ell = 1$, then $x \in C_1$ and $x \notin C_2+\cdots+C_n$. Hence, from the remark on how to choose C_1 in the composition series, we get $K\cdot C_i = 0$ $(2 \leqq i \leqq n)$ and $K\cdot C_1 = K\cdot D = eK^2 \geqq 2$.) In 8.2, if we put $h=1, k=0$

and $\Xi_i = \mathscr{O}(mK - Z_i - x)$, then

$$\Xi_{i+1}/\Xi_i \cong \mathscr{O}(F_{i+1} - \delta_{i\ell}x)_{C_i} \qquad (1 \leqq i \leqq n) \qquad (**)$$

holds. In fact, (i) for $i \leqq \ell - 1$, we have $\Xi_{i+1} \cong \mathscr{O}(F_{i+1})$, $\Xi_i \cong \mathscr{O}(F_{i+1} - C_i)$ and $\Xi_{i+1}/\Xi_i \cong \mathscr{O}(F_{i+1})_{C_i}$; (ii) for $i = \ell$, we have $\Xi_{\ell+1} \cong \mathscr{O}(F_{\ell+1} - x)$, $\Xi_\ell \cong \mathscr{O}(F_{\ell+1} - x - C_\ell)$ and $\Xi_{\ell+1}/\Xi_\ell \cong \mathscr{O}(F_{\ell+1} - x)_{C_\ell}$; (iii) for $i \geqq \ell + 1$, we have $\Xi_{i+1} \cong \mathscr{O}(F_{i+1} - x)$, $\Xi_i \cong \mathscr{O}(F_{i+1} - x - C_i)$ and $\Xi_{i+1}/\Xi_i \cong \mathscr{O}(F_{i+1} - x)_{C_i} = \mathscr{O}(F_{i+1})_{C_i}$. Hence (**) holds and we get $h_i = \delta_{i\ell}$ and $k_i = 0$. Since $m - e - 1 \geqq 1$ by hypothesis and $KC_\ell \geqq 1 + \delta_{\ell 1}$, we get

$$(m - e - 1)K \cdot C_i + D_{i-1} \cdot C_i \geqq 1 + \delta_{i\ell} > \frac{1}{4}(1 + \delta_{i\ell})^2 + \frac{1}{4}$$

by (α). We infer from 8.4 that $\dim H^1(S, \mathscr{O}((m - e)K)) \geqq \dim H^1(S, \Xi_{n+1})$. Since $m \geqq m_0$, however, we get $\dim H^1(S, \mathscr{O}((m - e)K)) = \dim H^1(S, \mathscr{O}(mK))$. Moreover, $\Xi_{n+1} = \mathscr{O}(mK - x)$ (8.2.1). In sum, we obtain $\dim H^1(S, \mathscr{O}(mK)) \geqq \dim H^1(S, \mathscr{O}(mK - x))$. We see from (*) that λ is an isomorphism and, hence, τ is surjective. This completes the proof for the case (I).

(II) $x \in \mathscr{E}$: We assume that $x \in \mathscr{E}_\lambda$. By 7.13, we can take a composition series $D = \sum_{i=1}^n C_i$ of D satisfying

$$C_n < \mathscr{E}_\lambda \quad \text{and} \quad (\beta)\ K \cdot C_i + D_{i-1} \cdot C_i \geqq 1.$$

Since $C_n < \mathscr{E}_\lambda < \mathscr{E}$, we get $K \cdot C_n = 0$. Therefore, K is trivial on C_n and $\mathscr{O}(mK)_{C_n} \cong \mathscr{O}_{C_n}$. We have the exact sequence

$$0 \to \mathscr{O}(mK - C_n) \to \mathscr{O}(mK) \to \mathscr{O}_{C_n} \to 0.$$

Since $C_n \cong \mathbb{P}^1$ and $H^1(C_n, \mathscr{O}_{C_n}) = 0$, we get

$$\begin{aligned} 0 \to H^0(S, \mathscr{O}(mK - C_n)) \to H^0(S, \mathscr{O}(mK)) &\xrightarrow{\tau'} H^0(C_n, \mathscr{O}_{C_n}) \\ \to H^1(S, \mathscr{O}(mK - C_n)) \xrightarrow{\sigma} H^1(S, \mathscr{O}(mK)) &\to 0. \end{aligned} \qquad (***)$$

In 8.2, we put $h = k = 0$ and $\Xi_i = \mathscr{O}(mK - Z_i)$. Therefore, we have $\Xi_{i+1}/\Xi_i \cong \mathscr{O}(F_{i+1})_{C_i}$ and $h_i = k_i = 0$ (8.3). By hypothesis, $m - e - 1 \geqq 1$. Since $(m - e - 1)K \cdot C_i + D_{i-1} \cdot C_i \geqq K \cdot C_i + D_{i-1} \cdot C_i \geqq 1 > \frac{1}{4} + \frac{1}{4}$, we infer from 8.4 that $\dim H^1(S, \mathscr{O}(m - e)K) \geqq \dim H^1(S, \Xi_n)$. On the other hand, since $m \geqq m_0$, we have $\dim H^1(S, \mathscr{O}(m - e)K) = \dim H^1(S, \mathscr{O}(mK))$ and $\Xi_n = \mathscr{O}(mK - C_n)$. Therefore, $\dim H^1(S, \mathscr{O}(mK)) \geqq \dim H^1(S, \mathscr{O}(mK - C_n))$. We see that σ is an isomorphism in (***) and, hence, τ' is surjective. Since K is trivial on C_n, if we take $\varphi \in H^0(S, \mathscr{O}(mK))$, then φ takes the constant value $\varphi(C_n)$ on C_n. In fact, we have $\tau' \colon \varphi \to \varphi(C_n)$. Moreover,

K is trivial on $\mathscr{E}_\lambda$ and φ as above is constant on $\mathscr{E}_\lambda$. Since $x \in \mathscr{E}_\lambda$, we have $\varphi(C_n) = \varphi(x)$. Therefore, returning to (**), we see that $\tau = \tau'$ and τ is surjective.

□

Theorem 8.7 $H^1(S, \mathscr{O}(mK)) = 0$ *for* $m \geqq 2$.

Proof Taking e sufficiently large, we have $P_e \geqq 2$ by the Riemann–Roch theorem, and $eK^2 \geqq 2$. If we put $F = (m-1)K$ and take an integer n satisfying $n(m-1) \geqq e+2+m$, then $|nF|$ has no base points by 8.6. On the other hand, $F^2 = (m-1)^2K^2 > 0$. Hence we infer from 6.3 that $H^1(S, \mathscr{O}(K+F)) = H^1(S, \mathscr{O}(mK)) = 0$. □

Corollary 8.8 P_m $(m \geqq 1)$ *is a topological invariant of* S. *In fact,*

$$\begin{cases} P_m = \frac{1}{2}m(m-1)c_1^2 + p_g - q + 1 & (m \geqq 2), \\ P_1 = p_g. \end{cases}$$

Proof 4.2 and 8.7. □

Summing up, we obtain:

Theorem 8.9 *If* $P_e \geqq 2$, $eK^2 \geqq 2$ *and* $m \geqq e+2$, *then the* m*-canonical system* $|mK|$ *has no base points. Therefore,* Φ_{mK} *is a holomorphic map.*

Proof By 8.7, we can take $m_0 = e+2$ in 8.5. Therefore, it is sufficient to assume $m \geqq e+2$ in order to assure the conditions on m in 8.6. □

8.10 We next consider whether Φ_{mK} is one-to-one (onto the image). For $\mathscr{E} = \mathscr{E}_1 + \cdots + \mathscr{E}_\lambda + \cdots + \mathscr{E}_r$ (7.6), since K is trivial on $\mathscr{E}_\lambda$, any $\varphi \in H^0(S, \mathscr{O}(mK))$ is constant on $\mathscr{E}_\lambda$. Hence $\Phi_{mK}(\mathscr{E}_\lambda)$ consists of one point, and Φ_{mK} fails to be one-to-one. Our answer to the problem will be given in 8.11.

Definition 8.10.1

(a) If $x \neq y$ and, if x and y are not contained in the same $\mathscr{E}_\lambda$, then we say that $x \neq y \bmod \mathscr{E}$ (x and y are different modulo $\mathscr{E}$).
(b) The holomorphic map $\Phi\colon S \to \mathbb{P}^n$ is said to be $1:1 \bmod \mathscr{E}$ (one-to-one modulo $\mathscr{E}$) if Φ satisfies the condition

$$\Phi^{-1}(\Phi(z)) = \begin{cases} z & (z \notin \mathscr{E}), \\ \mathscr{E}_\lambda & (z \in \mathscr{E}_\lambda). \end{cases}$$

$$(\Longleftrightarrow \text{If } x \neq y \bmod \mathscr{E}, \text{ then } \Phi(x) \neq \Phi(y).)$$

Remark 8.10.2 For $x \neq y$, if $H^1(S, \mathscr{O}(mK - x - y)) = 0$, then $\Phi_{mK}(x) \neq \Phi_{mK}(y)$.

Proof We have the following exact sequence (5.0 and 5.0.1):

$$0 \to \mathcal{O}(mK - x - y) \to \mathcal{O}(mK) \to \mathbb{C}_x \bigoplus \mathbb{C}_y \to 0.$$

Hence by hypothesis, we get the exact sequence

$$\cdots \to H^0(S, \mathcal{O}(mK)) \xrightarrow{\lambda} \mathbb{C}^2 \to 0,$$

where $\lambda\colon \varphi \to (\varphi(x), \varphi(y))$ for $\varphi \in H^0(S, \mathcal{O}(mK))$. Since λ is surjective, we can find $\varphi_0, \varphi_1 \in H^0(S, \mathcal{O}(mK))$ satisfying $(1, 0) = (\varphi_0(x), \varphi_0(y))$ and $(0, 1) = (\varphi_1(x), \varphi_1(y))$. We complete a basis $\{\varphi_0, \varphi_1, \dots\}$ for $H^0(S, \mathcal{O}(mK))$ by extending φ_0, φ_1. Then $\Phi_{mK} : z \to (\varphi_0(z), \varphi_1(z), \varphi_2(z), \dots) \in \mathbb{P}^n$ (modulo biholomorphic maps), and we have $\Phi_{mK}(x) = (1, 0, \dots)$ and $\Phi_{mK}(y) = (0, 1, \dots)$. Hence $\Phi_{mK}(x) \neq \Phi_{mK}(y)$. □

Theorem 8.11 *Assume that* $P_e \geqq 3$, $eK^2 \geqq 2$ *and* $m \geqq e + 3$. *Then the following hold*:

(a) *If* $x \neq y$ *mod* $\mathcal{E}$, *then* $H^1(S, \mathcal{O}(mK - x - y)) = 0$.
(b) Φ_{mK} *is holomorphic and one-to-one modulo* $\mathcal{E}$.

Proof If (a) is shown, then (b) follows from it, 8.9 and 8.10.2. We shall show (a). We separately consider three cases: (I) $x, y \notin \mathcal{E}$, (II) $x, y \in \mathcal{E}$, and (III) $x \notin \mathcal{E}$ but $y \in \mathcal{E}$.

(I) $x, y \notin \mathcal{E}$: Since $\dim H^0(S, \mathcal{O}(eK)) \geqq 3$, we can find $D \in |eK|$ satisfying $x, y \in D$ (which can be seen as follows: Letting $\{\varphi_0, \varphi_1, \varphi_2, \dots, \varphi_{P_e-1}\}$ be a basis for $H^0(S, \mathcal{O}(eK))$, we can find a solution $\varphi = \sum_{\nu=0}^{P_e-1} a_\nu \varphi_\nu$, $(a_0, a_1, a_2, \dots) \neq (0, 0, \dots, 0)$, of the simultaneous linear equation $\varphi(x) = \varphi(y) = 0$. We take such $\varphi \in H^0(S, \mathcal{O}(eK))$ and put $D = (\varphi)$.) We infer from 7.9 that there exists a composition series $D = \sum_{i=1}^n C_i$ of D which satisfies

$$(\alpha) \quad K \cdot C_1 \geqq 1, \quad D_{i-1} \cdot C_i \geqq 1 \qquad (i \geqq 2).$$

In particular, if we have a component Θ of D satisfying $x \notin \Theta$ and $k \cdot \Theta \geqq 1$, then we put $C_1 = \Theta$ and complete the composition series (7.9.1). In 8.2, we put $h = k = 1$ and $\Xi_i = \mathcal{O}(mK - Z_i - x - y)$. We assume that $x \in C_\ell$, $x \notin Z_{\ell+1}$, $y \in C_j$ and $y \notin Z_{j+1}$ in the series $D = \sum_{i=1}^n C_i$. We may assume that $\ell \leqq j$. Then,

$$\Xi_{i+1}/\Xi_i \cong \mathcal{O}(F_{i+1} - \delta_{i\ell}x - \delta_{ij}y)_{C_i} \qquad (1 \leqq i \leqq n). \qquad (*)$$

This can be seen as follows. If:

(i) $i \leqq \ell - 1$, then $\Xi_{i+1} \cong \mathcal{O}(F_{i+1})$, $\Xi_i(F_{i+1-C_i})$ and $\Xi_{i+1}/\Xi_i \cong \mathcal{O}(F_{i+1})_{C_i}$.
(ii) $i = \ell$, then $\Xi_{\ell+1} \cong \mathcal{O}(F_{\ell+1} - x)$, $\Xi_\ell \cong \mathcal{O}(F_{\ell+1} - C_\ell)$ and $\Xi_{\ell+1}/\Xi_\ell \cong \mathcal{O}(F_{\ell+1} - x)_{C_\ell}$ (since $x \in C_\ell$).
(iii) $\ell + 1 \leqq i \leqq j - 1$, then $\Xi_{i+1} \cong \mathcal{O}(F_{i+1} - x)$, $\Xi_i \cong \mathcal{O}(F_{i+1} - x - C_i)$ and $\Xi_{i+1}/\Xi_i \cong \mathcal{O}(F_{i+1} - x)_{C_i} = \mathcal{O}(F_{i+1})_{C_i}$ (since $x \notin C_i$).
(iv) $i = j$, then $\Xi_{j+1} \cong \mathcal{O}(F_{j+1} - y)$, $\Xi_j \cong \mathcal{O}(F_{j+1} - y - C_j)$ and $\Xi_{j+1}/\Xi_j \cong \mathcal{O}(F_{j+1} - y)_{C_j}$.
(v) $i \geqq j + 1$, then $\Xi_{i+1} \cong \mathcal{O}(F_{i+1} - x - y)$, $\Xi_i \cong \mathcal{O}(F_{i+1} - x - y - C_i)$ and $\Xi_{i+1}/\Xi_i \cong \mathcal{O}(F_{i+1} - x - y)_{C_i} = \mathcal{O}(F_{i+1})_{C_i}$ (since $x, y \notin C_i$).

Hence we get $(*)$.

Now, we put $(m - e - 1)K \cdot C_i + D_{i-1} \cdot C_i \geqq 2K \cdot C_i + D_{i-1} \cdot C_i =: r_i$. Since $C_\ell, C_j \nless \mathcal{E}$, we get $KC_\ell \geqq 1$ and $KC_j \geqq 1$. In particular, paying special attention to that we have $KC_\ell \geqq 1 + \delta_{\ell 1}$ as in 8.6 by the remark on the first curve of the composition series, we have: if $i = \ell = 1$, then $r_i \geqq 4$; if $i = \ell \geqq 2$ or $i = j > \ell$, then $r_i \geqq 3$ by (α); if $i \neq \ell, j$, then $r_i \geqq 1$ in view of (α). Therefore, we get $(m - e - 1)K \cdot C_i + D_{i-1} \cdot C_i \geqq 1 + \delta_{i\ell} + \delta_{ij}$. Since we have $h_i = \delta_{i\ell}$ and $k_i = \delta_{ij}$ by $(*)$ (see 8.3), we get

$$\frac{1}{4}(h_i + 1)^2 + \frac{1}{4}(k_i + 1)^2 = \frac{1}{2} + \frac{3}{4}\delta_{i\ell} + \frac{3}{4}\delta_{ij} < 1 + \delta_{i\ell} + \delta_{ij}.$$

Hence, we infer from 8.4 that $\dim H^1(S, \mathcal{O}((m - e)K)) \geqq \dim H^1(S, \Xi_{n+1})$. On the other hand, since $m - e \geqq 3$ by hypothesis, it follows from 8.7 that $H^1(S, \mathcal{O}(m - e)K) = 0$. Therefore, since we have $\Xi_{n+1} = \mathcal{O}(mK - x - y)$ (8.2.1), we obtain $H^1(mK - x - y) = 0$.

(II) $x, y \in \mathcal{E}$: We assume $x \in \mathcal{E}_\lambda$, $y \in \mathcal{E}_\nu$. Since $x \neq y \bmod \mathcal{E}$, we have $\lambda \neq \nu$. We take $D \in |eK|$ such that $x, y \in D$ (this is possible by $\dim |eK| \geqq 2$ (see (I) for example)). We infer from 7.12 that there is a composition series $D = \sum_{i=1}^n C_i$ of D satisfying

$$C_{n-1} < \mathcal{E}_\lambda, \quad C_n < \mathcal{E}_\nu \quad \text{and} \quad (\beta)\ K \cdot C_i + D_{i-1} \cdot C_i \geqq 1 \quad (1 \leqq i \leqq n).$$

In 8.2, we put $h = k = 0$ and $\Xi_i = \mathcal{O}(mK - Z_i)$. Then $\Xi_{i+1}/\Xi_i = \mathcal{O}(F_{i+1})_{C_i}$ and $h_i = k_i = 0$ for $1 \leqq i \leqq n$ (8.3). On the other hand, since

$$(m - e - 1)K \cdot C_i + D_{i-1} \cdot C_i \geqq K \cdot C_i + D_{i-1} \cdot C_i \geqq 1 > \frac{1}{4} + \frac{1}{4},$$

it follows from 8.4 that $\dim H^1(S, \mathcal{O}((m - e)K) \geqq \dim H^1(S, \Xi_{n-1})$. Since $m - e \geqq 3$, we have $H^1(S, \mathcal{O}((m - e)K)) = 0$ by 8.7 and, therefore, we get $H^1(S, \Xi_{n-1}) = 0$. On the other hand, since $K \cdot C_{n-1} = 0$ and $K \cdot C_n = 0$, we see that K is trivial on C_{n-1}, C_n and we have $\mathcal{O}(mK)_{C_{n-1}} \cong \mathcal{O}_{C_{n-1}}$,

$\mathcal{O}(mK)_{C_n} \cong \mathcal{O}_{C_n}$. Hence, we obtain the following exact sequence (note that $C_{n-1} \cap C_n = \emptyset$):

$$0 \to \mathcal{O}(mK - C_{n-1} - C_n) \to \mathcal{O}(mK) \to \mathcal{O}_{C_{n-1}} \bigoplus \mathcal{O}_{C_n} \to 0.$$

Since $\Xi_{n-1} = \mathcal{O}(mK - C_{n-1} - C_n)$, we get the exact sequence

$$0 \to H^0(S, \mathcal{O}(mK - C_{n-1} - C_n)) \to H^0(S, \mathcal{O}(mK)) \xrightarrow{\lambda} \mathbb{C} \bigoplus \mathbb{C} \to 0,$$

where $\lambda\colon \varphi \to (\varphi(C_{n-1}), \varphi(C_n))$, noting that K is trivial both on $\mathscr{E}_\lambda$ ($> C_{n-1}$) and $\mathscr{E}_\nu$ ($> C_n$), and $\varphi \in H^0(S, \mathcal{O}((mK)))$ takes the constant values $\varphi(C_{n-1})$, $\varphi(C_n)$ on $\mathscr{E}_\lambda$, $\mathscr{E}_\nu$, respectively. Now, since we have $x \in \mathscr{E}_\lambda$ and $y \in \mathscr{E}_\nu$, we see that $\varphi(C_{n-1}) = \varphi(x)$ and $\varphi(C_n) = \varphi(y)$. Hence $\lambda\colon \varphi \to (\varphi(x), \varphi(y))$. On the other hand, we have the following exact sequences:

$$0 \to \mathcal{O}(mK - x - y) \to \mathcal{O}(mK) \to \mathbb{C}_x \bigoplus \mathbb{C}_y \to 0.$$
$$0 \to H^0(S, \mathcal{O}(mK - x - y)) \to H^0(S, \mathcal{O}(mK)) \xrightarrow{\lambda'} \mathbb{C} \bigoplus \mathbb{C}$$
$$\to H^1(S, \mathcal{O}(mK - x - y)) \xrightarrow{\tau} H^1(S, \mathcal{O}(mK)) \to 0,$$

where $\lambda'\colon \varphi \to (\varphi(x), \varphi(y))$ for $\varphi \in H^0(S, \mathcal{O}(mK))$. Therefore, $\lambda' = \lambda$ and it is surjective. Hence we know that τ is an isomorphism. Since $H^1(S, \mathcal{O}(mK)) = 0$ by 8.7, we obtain $H^1(S, \mathcal{O}(mK - x - y)) = 0$.

(III) $x \in \mathscr{E}$, $y \notin \mathscr{E}$: This case can be done similarly to the above. Let $x \in \mathscr{E}_\lambda$ and choose a composition series $D = \sum_{i=1}^n C_i$ of D satisfying $C_n < \mathscr{E}_\lambda$ and (β) (7.13). Put $\Xi_i = \mathcal{O}(mK - Z_i - y)$. If $y \in C_\ell$ but $y \notin Z_{\ell+1}$, then $K \cdot C_\ell \geqq 1$ and $\Xi_i/\Xi_{i+1} \cong \mathcal{O}(F_{i+1} - \delta_{i\ell} y)_{C_i}$. We have $(m-e-1)K \cdot C_i + D_{i-1} \cdot C_i \geqq 1 + \delta_{i\ell}$ and, hence, $H^1(S, mK - C_n - y) = 0$ by 8.4 and 8.7. Then $0 \to H^0(S, mK - C_n - y) \to H^0(S, mK) \to H^0(C_n, \mathcal{O}_{C_n}) \bigoplus \mathbb{C} \to 0$ is exact. From this, we infer readily that the restriction map $H^0(S, mK) \to \mathbb{C}^2$ ($\varphi \to (\varphi(x), \varphi(y))$) is surjective, since $x \in \mathscr{E}_\lambda$ and $E_n < \mathscr{E}_\lambda$.

□

8.12 Next, we study whether Φ_{mK} is biholomorphic on $S \setminus \mathscr{E}$.

Since we know that Φ_{mK} is 1:1 on $S \setminus \mathscr{E}$ under the assumption of 8.11, it is sufficient to see that the rank of the Jacobian matrix for Φ_{mK} at each point $x \in S \setminus \mathscr{E}$ is always 2.

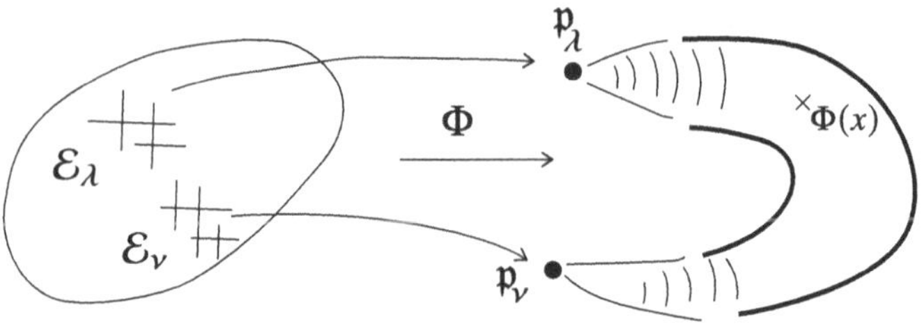

Remark 8.12.1 If $H^1(S, \mathcal{O}(mK - 2x)) = 0$, then Φ_{mK} is biholomorphic in a neighborhood of x.

Proof We have the following exact sequences (5.0 and 5.0.1 and the assumption):

$$0 \to \mathcal{O}(mK - 2x) \to \mathcal{O}(mK) \to \mathbb{C}^3_x \to 0,$$
$$\cdots \to H^0(S, \mathcal{O}(mK)) \xrightarrow{\lambda} \mathbb{C}^3 \to 0,$$

where, letting (z_1, z_2) be a system of local coordinates on S with the center x and writing $\psi \in H^0(S, \mathcal{O}(mK))$ as $\psi = \psi_0 + \psi_1 z_1 + \psi_2 z_2 + \psi_3 z_1^2 + \psi_4 z_1 z_2 + \psi_5 z_2^2 + \cdots$ in a neighborhood of x, we have $\lambda\colon \psi \to (\psi_0, \psi_1, \psi_2)$. Since λ is surjective, there are $\varphi_0, \varphi_1, \varphi_2 \in H^0(S, \mathcal{O}(mK))$ satisfying $(\varphi_{00}, \varphi_{01}, \varphi_{02}) = (1, 0, 0)$, $(\varphi_{10}, \varphi_{11}, \varphi_{12}) = (0, 1, 0)$ and $(\varphi_{20}, \varphi_{21}, \varphi_{22}) = (0, 0, 1)$. Completing the basis for $H^0(S, \mathcal{O}(mK))$ as $\{\varphi_0, \varphi_1, \varphi_2, \ldots\}$, we may assume that $\Phi_{mK}(z) = (\varphi_0(z), \varphi_1(z), \varphi_2(z), \ldots) = (1 + \cdots, z_1 + \cdots, z_2 + \cdots, \varphi_3(z), \ldots)$, where $\cdots$ in the first three components mean terms of order $\geqq 2$ in z_1, z_2. Therefore, the Jacobian matrix for Φ_{mK} is of rank 2 at x. □

Theorem 8.13 *If $P_e \geqq 4$, $eK^2 \geqq 2$ and $m \geqq e + 3$, then the following hold:*

(a) $H^1(S, \mathcal{O}(mK - 2x)) = 0$ *for* $x \notin \mathcal{E}$.
(b) Φ_{mK} *is holomorphic and biholomorphic modulo $\mathcal{E}$ (that is, biholomorphic on $S \setminus \mathcal{E}$). In particular, it is a birational holomorphic map.*

Proof If (a) is shown, then (b) follows from it, 8.9 and 8.12.1. We shall show (a). Since $\dim H^0(S, \mathcal{O}(eK)) = P_e \geqq 4$, we can find non-zero $\varphi \in H^0(S, \mathcal{O}(eK))$ satisfying $\varphi(x) = 0$, $(\frac{\partial \varphi}{\partial z_1})(x) = 0$ and $(\frac{\partial \varphi}{\partial z_2})(x) = 0$ (e.g., the argument in 8.11), where (z_1, z_2) denotes a system of local coordinates on S with the center x. We put $D = (\varphi) \in |eK|$. Note that x is a multiple point of D. We infer from 7.9 that there is a composition series $D = \sum_{i=1}^n C_i$ satisfying

$$(\alpha) \qquad K \cdot C_1 \geqq 1, \quad D_{i-1} \cdot C_i \geqq 1 \qquad (2 \leqq i \leqq n).$$

In particular, if there is a component Θ of D satisfying $x \notin \Theta$ and $K \cdot \Theta \geqq 1$, then we can take $C_1 = \Theta$ (7.9.1). In 8.2, put $h = 2$, $k = 0$ and $\Xi_i = \mathcal{O}(mK - Z_i - 2x)$. We assume that $x \in C_\ell$, $x \notin Z_{\ell+1}$ in $D = \sum_{i=1}^n C_i$. Then $K \cdot C_\ell \geqq 1 + \delta_{\ell 1}$ (by the same argument as in 8.6). We separately consider two cases: (I) x is a multiple point of C_ℓ, and (II) x is a simple point of C_ℓ.

If (I) is the case, then

$$\Xi_{i+1}/\Xi_i \cong \mathcal{O}(F_{i+1} - 2\delta_{i\ell}x)_{C_i}. \tag{*}$$

This can be seen as follows. If (i) $i \leqq \ell - 1$, then x is a multiple point of Z_i, Z_{i+1} and, hence, $\Xi_{i+1} \cong \mathcal{O}(F_{i+1})$, $\Xi_i \cong \mathcal{O}(F_{i+1} - C_i)$ and $\Xi_{i+1}/\Xi_i \cong \mathcal{O}(F_{i+1})_{C_i}$. If (ii) $i = \ell$, then $\Xi_{\ell+1} \cong \mathcal{O}(F_{\ell+1} - 2x)$, $\Xi_\ell \cong \mathcal{O}(F_{\ell+1} - 2x - C_\ell)$ and $\Xi_{\ell+1}/\Xi_\ell \cong \mathcal{O}(F_{\ell+1} - 2x)_{C_\ell}$. If (iii) $i \geqq \ell + 1$, then similarly $\Xi_{i+1}/\Xi_i \cong \mathcal{O}(F_{i+1} - 2x)_{C_i} = \mathcal{O}(F_{i+1})_{C_i}$ since $x \notin C_i$. Hence we get (*).

If we put $(m - e - 1)K \cdot C_i + D_{i-1} \cdot C_i \geqq 2K \cdot C_i + D_{i-1} \cdot C_i =: r_i$, then: if $i = \ell = 1$, then $r_i \geqq 4$ from $K \cdot C_\ell \geqq 1 + \delta_{\ell 1}$; if $i = \ell > 1$, then $r_i \geqq 3$ from $K \cdot C_\ell \geqq 1$ and (α); if $i \neq \ell$, then $r_i \geqq 1$ from (α). Therefore, we get $(m - e - 1)K \cdot C_i + D_{i-1} \cdot C_i \geqq 1 + 2\delta_{i\ell}$. Since $h_i = \delta_{i\ell}$ and $k_i = 0$ by ($*$) (8.3), we get $\frac{1}{4}(h_i + 1)^2 + \frac{1}{4}(k_i + 1)^2 = \frac{1}{2} + 2\delta_{i\ell} < 1 + 2\delta_{i\ell}$. Then we infer from 8.4 that $\dim H^1(S, \mathcal{O}((m - e)K)) \geqq \dim H^1(S, \Xi_{n+1})$. On the other hand, we have $H^1(S, \mathcal{O}((m - e)K)) = 0$ by 8.7 and $\Xi_{n+1} = \mathcal{O}(mK - 2x)$. It follows that $H^1(S, \mathcal{O}(mK - 2x)) = 0$ and we finish (I).

If (II) is the case, we can find another index $j < \ell$ with $x \in C_j$ and $x \notin C_{j+1} + \cdots + C_{\ell-1}$. In fact, since x is a simple point of C_ℓ, we can find such a C_j by the fact that x is a multiple point of D. We have

$$\Xi_{i+1}/\Xi_i \cong \mathcal{O}(F_{i+1} - (\delta_{ij} + 2\delta_{i\ell})x)_{C_i} \qquad (1 \leqq i \leqq n). \tag{**}$$

This can be seen as follows. If (i) $i \leqq j - 1$, then x is a multiple point of Z_{i+1}, Z_i and, hence, we have $\Xi_{i+1} = \mathcal{O}(mK - Z_{i+1})$, $\Xi_i = \mathcal{O}(mK - Z_{i+1} - C_i)$ and $\Xi_{i+1}/\Xi_i \cong \mathcal{O}(F_{i+1})_{C_i}$. If (ii) $i = j$, then x is a simple point of Z_{j+1} and $\Xi_{j+1} \cong \mathcal{O}(F_{j+1} - x)$. To see this, it suffices to show that $(\Xi_{j+1})_x \cong \mathcal{O}(F_{j+1} - x)_x$. For this purpose, we take a system of local coordinates (z_1, z_2) on S with the center x so that C_ℓ is defined by $z_2 = 0$ in a neighborhood of x.

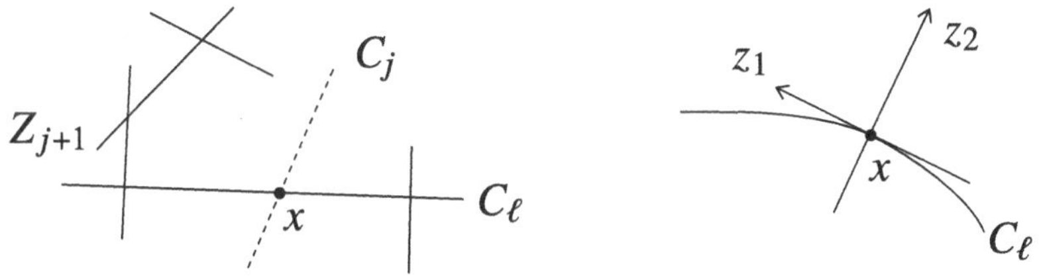

Then, the necessary and sufficient condition for $\varphi \in \mathcal{O}(mK - Z_{j+1})_x$ to be in $\mathcal{O}(mK - Z_{j+1} - 2x)_x$ is that φ can be expressed as $\varphi(z_1, z_2) = z_2\psi(z_1, z_2)$ with $\psi \in \mathcal{O}(F_{j+1})_x$ satisfying $\psi(0, 0) = 0$, that is, $\psi \in \mathcal{O}(F_{j+1} - x)_x$. Hence $\Xi_{j+1} = \mathcal{O}(mK - Z_{j+1} - 2x) \cong \mathcal{O}(F_{j+1} - x)$. On the other hand, since x is a multiple point of Z_j, we have $\Xi_j \cong \mathcal{O}(F_{j+1} - C_j) = \mathcal{O}(F_{j+1} - x - C_j)$. Therefore, $\Xi_{j+1}/\Xi_j \cong \mathcal{O}(F_{j+1} - x)$. If (iii) $j < i < \ell$, then it is not hard to see

$\Xi_{i+1}/\Xi_i \cong \mathcal{O}(F_{i+1})_{C_i}$. If (iv) $i = \ell$, then $\Xi_{\ell+1}/\Xi_\ell \cong \mathcal{O}(F_{\ell+1} - 2x)_{C_\ell}$. If (v) $i > \ell$, then $\Xi_{i+1}/\Xi_i \cong \mathcal{O}(F_{i+1})_{C_i}$. Hence we get (**).

If we put $(m - e - 1)K \cdot C_i + D_{i-1} \cdot C_i \geqq 2K \cdot C_i + D_{i-1} \cdot C_i =: r_i$, then, using $K \cdot C_j \geqq 1$, $K \cdot C_\ell \geqq 1$ (by $C_j, C_\ell \not< \mathscr{E}$) and ($\alpha$), we can show that $r_i \geqq 1$ when $i \neq j, \ell$; $r_i \geqq 2$ when $i = j$; $r_i \geqq 3$ when $i = \ell$. Therefore, $(m - e - 1)K \cdot C_i + D_{i-1} \cdot C_i \geqq 1 + \delta_{ij} + 2\delta_{i\ell}$. On the other hand, we infer from (**) that $h_i = \delta_{ij} + 2\delta_{i\ell}$, $k_i = 0$ (8.3). It follows that $\frac{1}{4}(h_i + 1)^2 + \frac{1}{4}(k_i + 1)^2 = \frac{1}{2} + \frac{3}{4}\delta_{ij} + 2\delta_{i\ell} < 1 + \delta_{ij} + 2\delta_{i\ell}$. Hence, as before, we use 8.4 and 8.7 to see $H^1(S, \mathcal{O}(mK - 2x)) = 0$. □

8.14 Finally, we consider what is the smallest possible value of e satisfying 8.9, 8.11, and 8.13 etc. Our answer will be found in 8.16.

For this purpose, we need the following lemma.

Lemma 8.15 *If $K^2 = 1$, then $p_g \leqq 2$ and $q \leqq 1$.*

Proof

(a) We shall show $p_g \leqq 2$. Assume that $p_g \geqq 2$. Since $\dim |K| = p_g - 1 \geqq 1$, we take a generic member $D \in |K|$ and write $D = C_1 + \cdot + C_n$ $(n \geqq 1)$. Since $1 = K^2 = K \cdot D = \sum_{i=1}^{n} KC_i$, we can assume that $K \cdot C_1 = 1$, $K \cdot C_i = 0$ $(i \geqq 2)$. Therefore, for $i \geqq 2$, we have $C_i < \mathscr{E}$ and $\pi(C_i) = 0$, $C_i^2 = -2$ (7.3). We claim that $n = 1$ and $D = C_1$. This can be seen as follows. If we put $X = \sum_{i=2}^{n} C_i$ $(\geqq 0)$, then $D = C_1 + X$ and X is the fixed part of $|K|$ and $C_1^2 \geqq 0$. We obtain $C_1^2 \geqq 1$ from $2\pi(C_1) - 2 = C_1^2 + K \cdot C_1$. Then $1 = K \cdot C_1 = D \cdot C_1 = C_1^2 + C_1X$, $C_1X \geqq 0$ and it follows that $C_1^2 = 1$, $C_1 \cdot X = 0$, which shows $X = 0$ in view of 7.8. Hence $D = C_1$. In this way, we know that a generic member of $|K|$ is an irreducible curve C with $C^2 = 1$. Then C has no singular points. (In fact, by Bertini's theorem, C does not have a singular point other than base points of $|K|$. On the other hand, choose any base point $\mathfrak{p}$ of $|K|$. If C' $(\neq C)$ is another generic member of $|K|$, then $\mathfrak{p} \in C'$. However, $C \cdot C' = C^2 = 1$ implies that $\mathfrak{p}$ is a simple point of C.) We have $\pi(C) = \frac{1}{2}(K \cdot C + C^2) + 1 = 2$. Since $K = [C]$, we have the following exact sequences:

$$0 \to \mathcal{O} \to \mathcal{O}(K) \to \mathcal{O}(K)_C \to 0,$$
$$0 \to H^0(S, \mathcal{O}) \to H^0(S, \mathcal{O}(K)) \to H^0(C, \mathcal{O}(K_C)) \to \cdots.$$

Since $H^0(S, \mathcal{O}) \cong \mathbb{C}$ and $\dim H^0(C, \mathcal{O}(K_C)) \leqq 1$, we get $p_g = \dim H^0(S, \mathcal{O}(K)) \leqq 2$. Hence $p_g = 2$.

(b) We shall show $q \leqq 1$: We have $q = K^2 + p_g - P_2 + 1$ (8.8). Since $K^2 = 1$ and $P_2 \geqq p_g$, we get $q \leqq 2$. We assume that $q = 2$ and derive a contradiction. Let φ_1, φ_2 be a basis for the space of holomorphic 1-forms on S. Then $d\varphi_\nu = 0$ $(\nu = 1, 2)$. Let $\{\gamma_1, \ldots, \gamma_4\}$ be a Betti basis for 1-cycles on S $(b_1(S) = 2q = 4)$. Put $\omega_{j\nu} = \int_{\gamma_j} \varphi_\nu$ $(1 \leqq j \leqq 4, 1 \leqq \nu \leqq 2)$, $\omega_j = (\omega_{j1}, \omega_{j2}) \in \mathbb{C}^2$,

$\Omega = \{\sum m_j \omega_j \mid m_j \in \mathbb{Z}\}$ and $\mathbf{A} = \mathbb{C}^2/\Omega$, and consider the Albanese map $\Psi \colon S \to \mathbf{A}$. We have $\Psi \colon z \to \Psi(z) = \left(\int_{p_0}^{z} \varphi_1, \int_{p_0}^{z} \varphi_2\right) \pmod{\Omega}$. We consider two cases: (1) $\varphi_1 \wedge \varphi_2 \neq 0$, (2) $\varphi_1 \wedge \varphi_2 = 0$.

(1) If $\varphi_1 \wedge \varphi_2 \neq 0$, then $\Psi(S) = \mathbf{A}$ and $D = (\varphi_1 \wedge \varphi_2) \in |K|$. As we did in (a), using $K \cdot D = K^2 = 1$, we can write $D = C + \sum_{i=2}^{n} E_i$, where $K \cdot C = 1$, $K \cdot E_i = 0$, $\pi(E_i) = 0$ and $E_i^2 = -2$. Then $\Psi(E_i)$ is a point. We claim that the restrictions φ_{1C}, φ_{2C} of φ_1, φ_2 to C are linearly independent over $\mathbb{C}$. [This can be seen as follows. If $(a_1\varphi_1 + a_2\varphi_2)_D = 0$ for $(a_1, a_2) \neq (0, 0)$, then $a_1 \int^z \varphi_1 + a_2 \int^z \varphi_2 =$ constant for $z \in C$ and we see that $\Psi(C) \subset \mathbf{A}$ is an elliptic curve. If we take a curve Γ satisfying $\Gamma \cap \Psi(C) = \emptyset$, $\Gamma \cap \Psi(E_i) = \emptyset$, then $\Psi^{-1}(\Gamma)$ does not meet $C + \sum E_i$ and, hence, $K \cdot \Psi^{-1}(\Gamma) = 0$. (Since $\Psi(C)$ is the image of a line in $\mathbb{C}^2$, if we take a line parallel to it, then we can get such Γ on $\mathbf{A}$.) This implies, however, that every component Θ_j of $\psi^{-1}(\Gamma)$ satisfies $K \cdot \Theta_j = 0$ and $\pi(\Theta_j) = 0$. Hence $\Psi(\Theta_j)$ is a point, which contradicts that $\Gamma = \bigcup_j \Psi(\Theta_j)$. We have shown that φ_{1C} and φ_{2C} are linearly independent.] On the other hand, C is non-singular and $\pi(C) = 2$, $D = C$. [Because: Let $\widetilde{C}$ be the non-singular model of C. By what we have just seen, there are two linearly independent holomorphic 1-forms $\tilde{\varphi}_1$ and $\tilde{\varphi}_2$ on $\widetilde{C}$. Thus $\pi(\widetilde{C}) \geqq 2$. On the other hand, $2 \leqq 2\pi(\widetilde{C}) - 2 \leqq 2\pi(C) - 2 = C^2 + K \cdot C = (D - \sum E_i) \cdot C + K \cdot C \leqq 2K \cdot C = 2$. This shows $\pi(C) = \pi(\widetilde{C}) = 2$ and, therefore, $C = \widetilde{C}$. Moreover, $C \cdot (\sum E_i) = 0$. Then, by 7.8, we get $\sum E_i = 0$ and conclude $D = C$.] Therefore, we get $(\varphi_1 \wedge \varphi_2) = C$. We have $\varphi_1(z) \neq 0$ when $z \notin C$. We remark that φ_{1C} has two zeros on C, since C is a non-singular curve with $\pi(C) = 2$. We can say the same thing for φ_{2C}. We may assume that they have two simple zeros and put $(\varphi_{1C}) = x + y$, $(\varphi_{2C}) = u + v$.

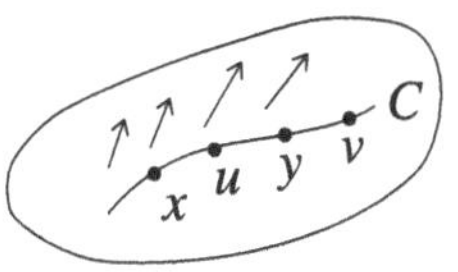

$$(\varphi_{1C}) = x + y$$

$$(\varphi_{2C}) = u + v$$

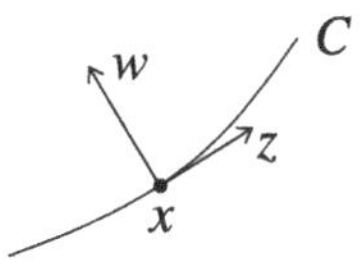

We regard the holomorphic 1-form φ_1 as a covariant vector field. Recall that c_2 is the Euler number of S. We use the formula $c_2 = \mathrm{I}_x(\varphi_1) + \mathrm{I}_y(\varphi_1)$, where $\mathrm{I}_x(\varphi_1)$, $\mathrm{I}_y(\varphi_1)$ denotes the algebraic index of the covariant vector field φ_1 at its singular point x, y. We can choose a system of local coordinates (w, z) with the center x on S so that $\mathrm{d}z = \varphi_2$, $\varphi_1 \wedge \varphi_2 = w\mathrm{d}w \wedge \mathrm{d}z$ (see figure above). Since then we have $(\varphi_1 - w\mathrm{d}w) \wedge \mathrm{d}z = 0$, we can write $\varphi_1 - w\mathrm{d}w = f\mathrm{d}z$. Since $\mathrm{d}\varphi_1 = 0$, however, we get $\mathrm{d}f \wedge \mathrm{d}z = 0$, that is, $\frac{\partial f}{\partial w} = 0$. This implies that f is a holomorphic function only in z and we can write $f = f(z)$. Consequently, we have $\varphi_{1C} = f(z)\mathrm{d}z = zg(z)\mathrm{d}z$ $(g(z) \neq 0)$ and, hence, $\varphi_1 = w\mathrm{d}w + gz\mathrm{d}z$ $(g \neq 0)$. Then we have $\mathrm{I}_x(\varphi_1) = 1$. Similarly, $\mathrm{I}_y(\varphi_1) = 1$. Therefore, from the above formula, we get $c_2 = 2$. Recall that Noether's formula 4.4 implies that $c_1^2 + c_2 = 12(p_g - q + 1) \equiv 0$ (12), which is impossible when $c_1^2 = K^2 = 1$, $c_2 = 2$.

(2) $\varphi_1 \wedge \varphi_2 = 0$: The Jacobian matrix for Ψ is of rank 1 everywhere, $\Psi(S) = \Delta \subset \mathbf{A}$ is a non-singular curve of genus 2, and $\Theta_u = \Psi^{-1}(u)$ is an irreducible non-singular curve for a general $u \in \Delta$. We have $p_g \geqq 1$. (This is because we have $12p_g = 7 + b_2$ from Noether's formula 4.4, $c_1^2 = K^2 = 1$, $c_2 = 2 - 4q + b_2$ and $q = 2$.) Then, from $\dim |K| = p_g - 1 \geqq 0$, we can take $|K| \ni D \geqq 0$. We have $D > 0$ by $K^2 = D^2 = 1$. If we write $D = \sum_{i=1}^{n} C_i$, then $1 = K^2 = K \cdot D = \sum_{i=1}^{n} K \cdot C_i$, $K \cdot C_i \geqq 0$. Hence we may assume that $K \cdot C_1 = 1$, $K \cdot C_i = 0$ $(i \geqq 2)$. We rewrite it as $D = C + \sum_{i=2}^{n} E_i$ as before (where $K \cdot C = 1$, $K \cdot E_i = 0$, $\pi(E_i) = 0$, $E_i^2 = -2$ (7.3)). Note that $\Psi(E_i)$ is one point ($\in \Delta$). Therefore, $\Theta_u \cap E_i = \emptyset$ (because, if $\Theta_u \cap E_i \neq \emptyset$, then $\Theta_u = E_i$. This contradicts that $\Theta_u^2 = 0$, $E_i^2 = -2$.) We have $K \cdot \Theta_u > 0$ by $\Theta_u^2 = 0$ and 7.3. Since $2\pi(\Theta_u) - 2 = K \cdot \Theta_u$, we get $K \cdot \Theta_u \geqq 2$. Hence $C \cdot \Theta_u = D \cdot \Theta_u = K \cdot \Theta_u \geqq 2$.

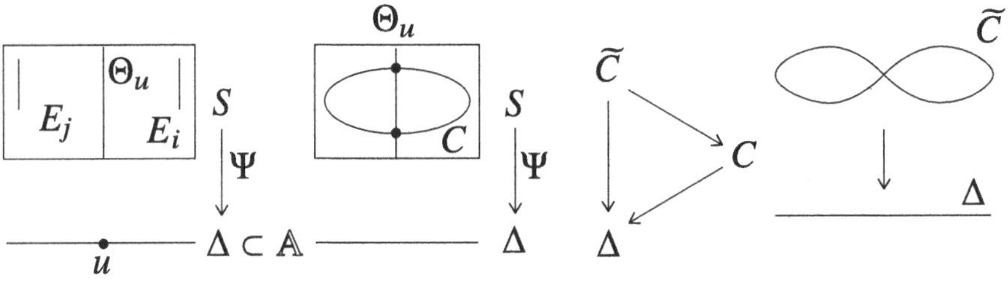

Let $\widetilde{C}$ denote the non-singular model of C. Then $\widetilde{C}$ is a (branched) covering of Δ. If we denote the covering degree by m and the degree of the ramification divisor by b, then $\chi(\widetilde{C}) = m\chi(\Delta) - b$ (where $\chi(*)$ denotes the Euler characteristic), that is, $2 - 2\pi(\widetilde{C}) = m(2 - 2\pi(\Delta)) - b$. Since $m = C \cdot \Theta_u \geqq 2$, we have $2\pi(\widetilde{C}) - 2 = m(2\pi(\Delta) - 2) + b \geqq 2m \geqq 4$ by $\pi(\Delta) = 2$. Hence $\pi(\widetilde{C}) \geqq 3$. On the other hand, we have

$$2\pi(C) - 2 = K \cdot C + C^2 = K \cdot C + C \cdot (D - \sum_i E_i)$$

$$= 2K \cdot C - \sum C \cdot E_i \qquad \text{(by } D \approx K\text{)}$$

$$\leqq 2K \cdot C = 2 \qquad \text{(by } CE_i \geqq 0\text{)},$$

which contradicts that $\pi(\widetilde{C}) \leqq \pi(C)$.

In both cases of (1) and (2), we are led to a contradiction. Therefore, $q \leqq 1$. □

Theorem 8.16 *$P_2 \geqq 2$ and $P_3 \geqq 4$.*

Proof We infer from 8.8 that $P_2 = K^2 + p_g - q + 1$ $(K^2 = c_1^2)$ and $P_3 = 3K^2 + p_g - q + 1 = 2K^2 + P_2$. Since we have $K^2 \geqq 1$ by Assumption 7.0, if we can show $P_2 \geqq 2$, then $P_3 \geqq 4$ follows. We shall show that $P_2 \geqq 2$. We consider the three cases $p_g \geqq 2$, $p_g = 1$, $p_g = 0$.

(a) $p_g \geqq 2$: Since $P_2 \geqq P_1 = p_g$, we get $P_2 \geqq 2$.
(b) $p_g = 1$: By Noether's formula 4.4, we have $q \leqq 2$ (by $p_a = p_g - q$, $K^2 = c_1^2$, $c_2 = b_2$). We have $K^2 - q \geqq 0$ by 8.15, and $P_2 = p_g + 1 + K^2 - q \geqq 2$.
(c) $p_g = 0$: We have $q = 0$ (Enriques). Hence $P_2 = K^2 + 1 \geqq 2$.

□

Theorem 8.17

(a) *If $m \geqq 4$, then Φ_{mK} is a holomorphic map.*
(b) *If $m \geqq 6$, then Φ_{mK} is biholomorphic modulo $\mathscr{E}$ (i.e., biholomorphic on $S \setminus \mathscr{E}$) and, therefore, it is a birational holomorphic map.*

Proof In view of 8.16, (a) follows from 8.9 and, (b) follows from 8.13. □

Bibliography

1. Brieskorn, E.: Über die Auflösung gewisser Singularitäten von holomorphen Abbildungen. Math. Ann. **166**, 76–102 (1966)
2. Enriques, F.: Le Superficie Algebriche. Nicola Zanichelli, Bologna (1949)
3. Gorenstein, D.: An arithmetic theory of adjoint plane curves. Trans. Am. Math. Soc. **12**, 414–436 (1952)
4. Kodaira, K.: On a differential-geometric method in the theory of analytic stacks. Proc. Nat. Acad. Sci. U.S.A. **39**, 1268–1273 (1953)
5. Mumford, D.: Appendix to [Zariski, O.: Ann. Math. **76**, 560–615 (1962)]
6. Mumford, D.: Pathologies III. Am. J. Math. **89**, 94–104 (1967)
7. Rosenlicht, M.: Equivalence relations on algebraic curves. Ann. Math. **56**, 169–191 (1952)
8. Šafarevič, I.R.: Algebraic surfaces. Proceedings of the Steklov Institute of Mathematics, vol. 75. American Mathematical Society, Providence (1965, 1967)
9. Zariski, O.: The theorem of Riemann–Roch for high multiples of an effective divisor on an algebraic surface. Ann. Math. **76**, 560–615 (1962)

K. Kodaira, *Theory of Algebraic Surfaces*, SpringerBriefs in Mathematics,
https://doi.org/10.1007/978-981-15-7380-4

Printed in the United States
By Bookmasters